Series Editor's Foreword

Oxford Chemistry Primers are designed to give a concise introduction to all chemistry students by providing the material that would usually form an 8–10 lecture course. As well as providing up-to-date information, this series expresses the explanations and rationales that form the framework of current understanding of inorganic chemistry.

Jon McCleverty has provided an essential guide to first-row transition element chemistry, which provides a fine precis of the theory and underlying structures and reactivity of this important part of the Periodic table. The quality and depth of this survey emanates from Jon's own research life at the heart of this field. Companion books covering the heavy transition elements and the f-block elements will succeed this volume. 3d-element chemistry is universal in undergraduate chemistry courses, and this text is perhaps the most approachable companion to these courses.

John Evans
Department of Chemistry,
University of Southampton

Preface

This short book is intended for students who have had at least one year's basic introduction to the concepts of transition metal chemistry. It is based on several excellent earlier Primers, overlapping slightly with and extending their exposé of this fascinating area of inorganic chemistry. Parts of this book are also intended as a taster for later, deeper studies.

Where possible, I have tried to show where a knowledge of structure and theory can amplify and explain experimental observation (the facts!). Inevitably, in a text of this kind, there will be omissions and mistakes. For the former I apologise by excusing myself on the grounds of limitation in length (96 pages in total, including the preface, index, etc.). For the latter I take full responsibility.

I would like to thank Mike Ward for some very useful general discussions, and I am especially grateful to Rowena Paul and Zoe Reeves for detailed criticisms and comments, both scientific and grammatical.

Now I truly understand the humble dedications to long-suffering families in the prefaces of so many textbooks. My wife Dianne has shown almost superhuman patience as this work was completed, and for that I am profoundly grateful.

Bristol   J. A. McC.
July 1998

Contents

Chemistry of the First-row Transition Metals

Jon McCleverty

Professor of Inorganic Chemistry, School of Chemistry
University of Bristol

Series sponsor: **ZENECA**

ZENECA is a major international company active in four main areas of business: Pharmaceuticals, Agrochemicals and Seeds, Specialty Chemicals, and Biological Products.

ZENECA's skill and innovative ideas in organic chemistry and bioscience create products and services which improve the world's health, nutrition, environment, and quality of life.

ZENECA is committed to the support of education in chemistry and chemical engineering.

OXFORD
UNIVERSITY PRESS

OXFORD

UNIVERSITY PRESS

Great Clarendon Street, Oxford OX2 6DP
Oxford University Press is a department of the University of Oxford
and furthers the University's aim of excellence in research, scholarship,
and education by publishing worldwide in

Oxford New York

Athens Auckland Bangkok Bogotá Buenos Aires Calcutta
Cape Town Chennai Dar es Salaam Delhi Florence Hong Kong Istanbul
Karachi Kuala Lumpur Madrid Melbourne Mexico City Mumbai
Nairobi Paris São Paulo Singapore Taipei Tokyo Toronto Warsaw

and associated companies in Berlin Ibadan

Oxford is a registered trade mark of Oxford University Press

Published in the United States
by Oxford University Press Inc., New York

© Jon McCleverty, 1999

British Library Cataloguing in Publication Data
Data available

Library of Congress Cataloging in Publication Data

ISBN 0–19–850151 X (Pbk)

Typeset by the author

Printed in Great Britain
on acid-free paper by
The Bath Press, Avon

1   Introduction

The Periodic Table of the elements is shown in Table 1.1. The *transition elements* can be defined strictly as those that, as elements, have partly filled *d* or *f* shells. This encompasses Groups 3 to 11, lying between the *s*-block elements (Groups 1 and 2) and the *p*-block elements and the noble gases, Groups 13 – 18. The *f*-block elements are made up of the lanthanoids and the actinoids. Elements 104–109, the post-actinoid elements, exist only in atom quantities but should have properties consistent with the *d*-block elements (their names and element symbols are provisional). This leaves Group 12 (Zn, Cd and Hg) in a kind of limbo, being neither transition elements nor usually regarded in the same category as the *p*-block or 'main group elements'. The position of hydrogen is ambivalent because of its unique properties.

Table 1.1 The Periodic Table

1	2	3	4	5	6	7	8	9	10	11	12	13	14	15	16	17	18
s block		*d* block										*p* block					
1 H																	2 He
3 Li	4 Be											5 B	6 C	7 N	8 O	9 F	10 Ne
11 Na	12 Mg											13 Al	14 Si	15 P	16 S	17 Cl	18 Ar
19 K	20 Ca	21 Sc	22 Ti	23 V	24 Cr	25 Mn	26 Fe	27 Co	28 Ni	29 Cu	30 Zn	31 Ga	32 Ge	33 As	34 Se	35 Br	36 Kr
37 Rb	38 Sr	39 Y	40 Zr	41 Nb	42 Mo	43 Tc	44 Ru	45 Rh	46 Pd	47 Ag	48 Cd	49 In	50 Sn	51 Sb	52 Te	53 I	54 Xe
55 Cs	56 Ba	57 La*	72 Hf	73 Ta	74 W	75 Re	76 Os	77 Ir	78 Pt	79 Au	80 Hg	81 Tl	82 Pb	83 Bi	84 Po	85 At	86 Rn
87 Fr	88 Ra	89 Ac†	104 Rf	105 Db	106 Sg	107 Bh	108 Hs	109 Mt									

								f block							
*Lanthanoids		58 Ce	59 Pr	60 Nd	61 Pm	62 Sm	63 Eu	64 Gd	65 Tb	66 Dy	67 Ho	68 Er	69 Tm	70 Yb	71 Lu
†Actinoids		90 Th	91 Pa	92 U	93 Np	94 Pu	95 Am	96 Cm	97 Bk	98 Cf	99 Es	100 Fm	101 Md	102 No	103 Lr

The elements of Groups 1 to 12, all of Group 13 except boron, most of Group 14 except carbon and silicon, Sb and Bi in Group 15 and Po in Group 16, are *metallic*. This means that they are normally lustrous, malleable solids capable of conducting heat and electricity, are electropositive, and normally form basic oxides and hydroxides. However, if the physical and chemical trends of the elements are examined more closely as we proceed from left to right across the Periodic Table, subtle but significant changes in these trends can be detected, particularly in passing from Group 4 through 11 to 12.

The transition elements have a number of properties in common which can be related to the presence of partly filled *d* orbitals: (a) they are all metals and almost all are hard, strong, high-melting, high-boiling, and conduct heat and electricity effectively; (b) they form alloys with one another and with other metallic elements; (c) many of them are electropositive enough to dissolve in mineral acids; (d) they exhibit a variety of oxidation states which may interchange by one unit, a significant difference from many of the *p*-block metals whose principal oxidation states usually differ by two units; (e) many of their compounds are coloured, a property mainly due to electronic transitions involving the *d* orbitals; and (f) because of partially filled *d* orbitals, many of their compounds are paramagnetic.

While the *s*-block elements readily form ionic compounds, those metallic elements from Groups 13 to 16 show a very high degree of covalent bonding. The transition from predominantly ionic to covalent bonding appears noticeable and becomes increasingly significant in progressing from Group 4 to 11. This is the main reason why this collection of elements is described as the *transition* elements.

The first row of the transition elements formally begins with scandium whose electronic configuration is $1s^22s^23s^23p^63d^14s^2$ and terminates with zinc, $1s^22s^23s^23p^63d^{10}4s^2$, *i.e.*, the 3*d* shell is filled. The second and third rows are associated with the filling of the 4*d* and 5*d* shells, and the bottom (fourth) row is made up of elements 104 to 111, representing the filling of the 6*d* shell.

1.1 The first-row transition elements

This text is concerned only with the first-row transition elements. Two questions arise immediately: (1) which elements should be considered as transition elements in this text, and (2) why consider the first row separately from their second-, third- and fourth-row congeners?

In addressing (1), we can see from the Periodic Table above that the first row formally begins with scandium and ends with zinc. However, despite the fact that the outer electron configuration of the former, *as an element*, is $4s^23d^1$ consistent with the definition of a transition element given above, scandium is almost never treated as such since it lacks one of the main characteristics of those metals: variable oxidation states. At the other end of the row, zinc has the outer electron configuration $3d^{10}4s^2$ and forms no compounds in which the 3*d* shell population is reduced from 10 electrons. So discussion of first-row transition element chemistry is usually confined to the metals Ti, V, Cr, Mn, Fe, Co, Ni and Cu, highlighted in Table 1.

In answering (2), there are two main reasons for considering the first-row transition elements apart from their heavier congeners:
(a) in each group (*e.g.*, Cr, Mo, and W, or Fe, Ru, and Os) the properties of the first-row element always differ significantly from the heavier elements, and so comparisons within each Group are of limited value;
(b) the aqueous chemistry of the first-row elements is relatively simple, and the use of crystal field theories (see below) in explaining many of the physical and chemical properties of compounds of these metals has been fairly successful compared with the heavier transition elements.

Electronic configuration and oxidation states

The energies of the 3*d* and 4*s* orbitals in the neutral atoms are quite similar so that while most configurations are of the $3d^n4s^2$ type, the exchange-energy stabilisation of the filled and half-filled shells gives $3d^54s^1$ for Cr and $3d^{10}4s^1$ for Cu.

When the atoms are ionised, the 3*d* orbitals become significantly more stabilised so that the 4*s* orbital and the ions all have $3d^n$ configurations. The configurations for the elements in oxidation states 0, II and III are shown in Table 1.2.

The relative difficulty in obtaining compounds containing Ni^{III} and Cu^{III} is due in large part to the relatively high values of the third ionisation enthalpies (Table 1.3). Because of the stability of the filled $3d$ shell, copper is the only element in this group which affords compounds containing Cu^{I} unsupported by π-acceptor ligands (Section 5.10, p.78).

Table 1.2 Electronic configuration of the first-row elements in oxidation states 0, II and III ([Ar] is the inner 'argon' core, $1s^2 2s^2 2p^6 3s^2 3p^6$)

M	M^0	M^{II}	M^{III}
Ti	$[Ar]3d^2 4s^2$	$[Ar]3d^2 4s^0$	$[Ar]3d^1 4s^0$
V	$[Ar]3d^3 4s^2$	$[Ar]3d^3 4s^0$	$[Ar]3d^2 4s^0$
Cr	$[Ar]3d^5 4s^1$	$[Ar]3d^4 4s^0$	$[Ar]3d^3 4s^0$
Mn	$[Ar]3d^5 4s^2$	$[Ar]3d^5 4s^0$	$[Ar]3d^4 4s^0$
Fe	$[Ar]3d^6 4s^2$	$[Ar]3d^6 4s^0$	$[Ar]3d^5 4s^0$
Co	$[Ar]3d^7 4s^2$	$[Ar]3d^7 4s^0$	$[Ar]3d^6 4s^0$
Ni	$[Ar]3d^8 4s^2$	$[Ar]3d^8 4s^0$	$[Ar]3d^7 4s^0$
Cu	$[Ar]3d^{10} 4s^1$	$[Ar]3d^9 4s^0$	$[Ar]3d^8 4s^0$

We must realise that although ionisation enthalpies provide some indication of the relative stabilities of oxidation states, we have to be careful to state the precise conditions under which the oxidation state is being defined. For example, low oxidation states may be perfectly stable in the absence of air or in fused melts, but have no real existence in aqueous media.

So, in defining the relative stability of oxidation states for a particular metal ion, it is necessary to include not only the ionisation enthalpies of the metal, but the ionic radius of the ion, its electronic structure, the nature of the ligands associated with the ion (polarisability, π-donor or π-acceptor properties), the stereochemistry of the ion either as a coordination compound (complex) or in a solid lattice, and the nature of solvents or other media supporting the ion.

There are some detectable trends in oxidation state stability for the first-row transition elements.

(a) From Ti to Mn the highest oxidation state is commensurate with the loss of all the $3d$ and $4s$ electrons. The stability of the highest state decreases from Ti^{IV} to Mn^{VII}, and after Mn the highest oxidation states are rarely encountered (Chapters 3 and 4).

(b) The oxides of the early first-row metals (Ti to Mn) become more acidic with increasing oxidation state, and the halides more covalent and prone to hydrolysis by water.

(c) The metal atoms in the oxoanions of oxidation states VI to VII are tetrahedrally coordinated by oxygen atoms, whereas in the oxides of metals with oxidation states lower than IV the atoms are usually octahedrally coordinated (Chapters 2 and 4).

(d) In the II and III oxidation states, complexes in aqueous solution are usually either four- or six-coordinate and are generally similar with respect to their chemical properties and reaction stoichiometries.

Table 1.3 Ionisation enthalpies of first-row transition elements, ΔH^0_{ion} (in kJ mol^{-1})

	1st	2nd	3rd
Ti	656	1309	2650
V	650	1414	2828
Cr	653	1592	3056
Mn	717	1509	3251
Fe	762	1561	2956
Co	758	1644	2131
Ni	737	1752	3489
Cu	745	1958	3545

ΔH^0_{ion} (n) for process $M(g) \rightarrow M^+(g) + e^-$, ΔH^0_{ion} (1), *etc.*

(e)   Oxidation states lower than II, with the notable exception of Cu^I, are found only in compounds containing ligands capable of π-acceptor behaviour (*e.g.*, CO, CNR, NO, PR_3, C_5H_5, alkenes, alkynes, etc.) (Chapter 5).

General comments on oxidation states –I to VII

I, 0, –I and lower. All first-row elements form some compounds in these states, but only with ligands with π-acceptor capability. An exception to this general rule is Cu^I.

II. All the elements form well-defined binary compounds in this state, such as oxides and halides, in which the bonding is predominantly ionic. With the exception of Ti, the metals form well-defined hexa-aqua ions, $[M(H_2O)_6]^{2+}$, and they also form a wide range of complexes.

III.   All the elements form at least some compounds in this oxidation state, which is the highest known for copper. The bonding in the fluorides and oxides is essentially ionic, although some chlorides, such as $FeCl_3$, have covalent character. The elements from Ti to Co form hexa-aqua ions, $[M(H_2O)_6]^{3+}$ although the Co^{III} and Mn^{III} species are strongly oxidising (*i.e.*, they are easily reduced, Section 3.2). In aqueous solution certain anions readily form complex species, *e.g.*, Fe^{III}. Because of their inertness to substitution, Cr^{III} and Co^{III} have an extensive chemistry (Chapter 4).

IV.   This is the most important oxidation state of Ti, and the main chemistry of this element is that of TiO_2 (Chapter 2) and $TiCl_4$ and compounds derived from it (Chapter 4). This state is also important in vanadium chemistry, being typified by the vanadyl ion, VO^{2+}. For the elements Cr to Ni, the IV state is found mainly in fluorides or complex fluoro anions, although there are some important oxo species.

V, VI and VII.   These states only occur in fluorides and oxofluorides of chromium (V – VI) and manganese (V – VII), *e.g.*, CrF_5, MnO_3F, and in the oxo anions $[M^nO_4]^{(8-n)-}$, *e.g.*, $[MnO_4]^-$. Compounds of these elements in these very high oxidation states are powerful oxidising agents (Section 3.2).

1.2   Electronic configuration–structure relationships

The $3d$ orbitals play *the* major role in the electronic structure of first-row transition elements, thereby determining in large part the structures of transition metal compounds (as complexes, binary and ternary compounds, *etc.*), their magnetic properties, their colour, and, frequently, their reactivity. The relationship between the *d*-orbital configuration and these physical and chemical properties can be rationalised by the so-called *ligand field theory*. This theory was derived from an electrostatic model, known usually as Crystal Field Theory (CFT), but modified by the recognition that most metal–ligand interactions involve varying degrees of covalency. The full development of ligand field theory is beyond the scope of this Primer, and most of the explanations and rationalisations used here will be derived mainly from CFT approximations. It will be assumed that the reader has encountered basic crystal field theory, so only a short résumé of its applications to aspects of first-row transition metal chemistry will be given here.

> For a description of basic crystal field theory, consult M. J. Winter, *d-Block chemistry*, Oxford Chemistry Primer 27, 1994.

Simple crystal field theory

The theory is based on what happens to the energy levels of free metal ions when there are placed in the electrostatic field caused by anions in a crystal, the so-called 'crystal field'.

The starting assumptions may be summarised as follows:

*   The central metal ion is regarded as a positively charged point surrounded by a set of negatively charged points representing the ligands (for the purposes of the assumptions it makes no difference whether the ligands are neutral or anionic).
*   The effect of any electrostatic field generated by the ligands will be to repel the electrons associated with the central metal ion. The consequence of this is to raise the energy of the d-orbital shell relative to that in the absence of the perturbing electrostatic field (the so-called free metal ion).
*   If the point negative charges (the ligands) are arranged in specific geometries (most frequently octahedral and tetrahedral), the relative energies of the d orbitals will be reorganised in order to conserve overall energy. This is a consequence of the differences in the shapes of the d orbitals (Fig. 1.1) and their spatial arrangements.
*   The electrostatic field is generally referred to as the *crystal field*.

Splitting of the 3d orbitals in six-coordination: the octahedral crystal field

In the great majority of first-row transition metal compounds the central metal atom is surrounded by six ligands in an octahedral array. The splitting of the $3d$ orbitals within the octahedral crystal field is then as shown in Fig. 1.2(a). This figure expresses the splitting energies in terms of the Δ_{oct} or $10Dq$, the t_{2g} set being relatively stabilised by $-0.4\Delta_{oct}$ or $-4Dq$ while the e_g set is destabilised by $-0.6\Delta_{oct}$ or $-6Dq$.

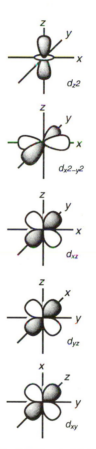

Fig. 1.1 The d orbitals

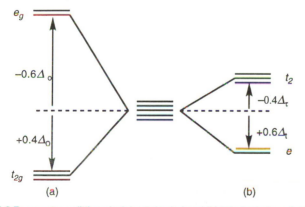

(a)   (b)

Fig.1.2 Barycentre splittings in (a) octahedral and (b) tetrahedral crystal fields

> The labels e_g and t_{2g} refer to doubly and triply degenerate orbitals in octahedral symmetry, and e and t_2 to doubly and triply degenerate orbitals in tetrahedral symmetry: the origin of the notation arises in Group Theory and need not concern us in this text.

$\Delta_{oct} = 10Dq$ = octahedral crystal field splitting energy.
$\Delta_{tet} = 10Dq'$ = tetrahedral crystal field splitting energy.

Splitting of the 3d orbitals in four-coordination: the tetrahedral crystal field

The splitting of the $3d$ orbitals within the tetrahedral crystal field is expressed as Δ_{tet} or $10Dq'$ (Fig. 1.2(b)). This can be somewhat confusing, since the magnitude of the effect for a tetrahedral species is significantly smaller than that for the octahedron for a comparable set of ligands and the same metal. The reason for the reduction in crystal field splitting energy is relatively easy to understand since, in the octahedron, six ligands give rise to the effect while in the tetrahedron, this is reduced to four. Furthermore, in the octahedron, two of the orbitals point directly at the ligands, while none of the orbitals in the tetrahedron do so. The relationship between the two splitting energies is given by the expression $\Delta_{tet} = 4/9\Delta_{oct}$.

Distortion of the octahedron towards tetragonal and then planar symmetry

As the ligand point charges are gradually removed along the z axis the octahedron distorts tetragonally and, on their ultimate loss, the remaining charges define a square plane.

So, as the ligands along the z axes are progressively removed, the repulsion felt by electrons in the orbitals having a z component, *viz.*, $3d_{z^2}$, $3d_{xz}$ and $3d_{yz}$, will be reduced, leading to a loss of the degeneracy of the e_g and t_{2g} levels. Thus, the e_g orbitals split into their components, and the t_{2g} levels split into a lower energy pair, $3d_{xz}$ and $3d_{yz}$, and a higher lying $3d_{xy}$ (Fig. 1.3).

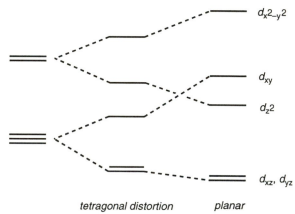

tetragonal distortion   planar

Fig. 1.3 The effect of partial and complete removal of point charges on the z axis of an octahedral crystal field

Final removal of the point charges on the z axes leads to an accentuation of the orbital stabilisation, also as shown in Fig. 1.3.

3d-electron configuration and crystal field stabilisation energy

Having established how the $3d$ orbitals split in the crystal fields most likely to be experienced by first-row transition metal compounds, the arrangement of electrons in these d orbitals can be considered. The progressive addition of electrons must take account of *Hund's rule* concerning electron spin multiplicity and the magnitude of Δ_{oct} (but not Δ_{tet}).

The first three electrons in an octahedral field are added sequentially to the t_{2g} orbitals, giving $3d^1$ or t_{2g}^1, $3d^2$ or t_{2g}^2, and $3d^3$ or t_{2g}^3, having one, two and three unpaired spins, respectively. The *crystal field stabilisation energy* (CFSE) may be calculated by inspection of Fig. 1.3. So the CFSE for a $3d^1$ configuration is either $-0.4\Delta_{oct}$ or $-4Dq$ and for $3d^2$, it is $-0.8\Delta_{oct}$ or $-8Dq$. It is not usually necessary to quote the value of these energies in kJ^{-1}, although that can be done. Units of Δ_{oct} or $10Dq$ are often enough.

For four d electrons, there are two possible configurations depending on the size of Δ_{oct} and the spin pairing energy, P. One has maximum spin multiplicity, $(t_{2g})^3(e_g)^1$, and the other will have two paired electrons, $(t_{2g})^4(e_g)^0$. The CFSEs will be $-0.6\Delta_{oct}$ or $-6Dq$ for the high-spin situation and $-1.6\Delta_{oct}$ or $-16Dq$ for the low-spin or spin-paired situation.

Electronic configurations for high- and low-spin configurations are summarised in Table 1.4.

> Pairing energy P is composed of two terms: an inherent repulsion term which must be overcome when forcing two electrons to occupy the same orbital, and the loss of exchange energy which occurs as electrons with parallel spins are forced to have antiparallel spins. The greatest loss of exchange energy occurs when the d^5 configuration is forced to pair.

Table 1.4 Configurations, spin multiplicity and CFSE for octahedral systems

$3d^n$	Spin multiplicity	u.e.[a]	Electronic configuration	CFSE
$3d^1$		1	$t_{2g}^1 e_g^0$	$-0.4\Delta_{oct}; -4Dq_o$
$3d^2$		2	$t_{2g}^2 e_g^0$	$-0.8\Delta_{oct}; -8Dq_o$
$3d^3$		3	$t_{2g}^3 e_g^0$	$-1.2\Delta_{oct}; -12Dq_o$
$3d^4$	high spin	4	$t_{2g}^3 e_g^1$	$-0.6\Delta_{oct}; -6Dq_o$
	low spin	2	$t_{2g}^4 e_g^0$	$-1.6\Delta_{oct} + P; -16Dq_o + P$
$3d^5$	high spin	5	$t_{2g}^3 e_g^2$	$0; 0$
	low spin	1	$t_{2g}^5 e_g^0$	$-2.0\Delta_{oct} + 2P; -20Dq_o + 2P$
$3d^6$	high spin	4	$t_{2g}^4 e_g^2$	$-0.4\Delta_{oct}; -4Dq_o$
	low spin	0	$t_{2g}^6 e_g^0$	$-2.4\Delta_{oct} + 2P; -24Dq_o + 2P$
$3d^7$	high spin	3	$t_{2g}^5 e_g^2$	$-0.8\Delta_{oct}; -8Dq_o$
	low spin	1	$t_{2g}^6 e_g^1$	$-1.8\Delta_{oct} + P; -18Dq_o + P$
$3d^8$		2	$t_{2g}^6 e_g^2$	$-1.2\Delta_{oct}; -12Dq_o$
$3d^9$		1	$t_{2g}^6 e_g^3$	$-0.6\Delta_{oct}; -6Dq_o$

[a] unpaired electrons.

There are three configurations for which there is no net CFSE: $3d^0$, high spin $3d^5$, and $3d^{10}$, which correspond to highly symmetrical arrangements of the empty, half-filled and completely filled $3d$ shell.

The CFSEs for the tetrahedral case can be worked out similarly, but it should be remembered that only the high-spin situation applies (there are no examples of low-spin tetrahedral compounds of first-row transition metals). The results are shown in Table 1.5.

When working out whether P should be included in CFSE, we must remember to compare the proposed configuration under the influence of the crystal field with the number of unpaired electrons in a spherical field. For the d^8 configuration there will be two unpaired electrons whether or not the crystal field is spherical, octahedral or tetrahedral: spin pairing energy is irrelevant to CFSE. For high spin, $\Delta o\chi\tau < P$ and for low spin, $\Delta o\chi\tau > P$.

Table 1.5 Configurations and CFSEs for tetrahedral systems

$3d^n$	u.e.[a]	Electronic configuration	CFSE
$3d^1$	1	$e^1 t_2^0$	$-0.6\Delta_{tet}$; $-6Dq_t$
$3d^2$	2	$e^2 t_2^0$	$-1.2\Delta_{tet}$; $-12Dq_t$
$3d^3$	3	$e^2 t_2^1$	$-0.8\Delta_{tet}$; $-8Dq_t$
$3d^4$	4	$e^2 t_2^2$	$-0.4\Delta_{tet}$; $-4Dq_t$
$3d^5$	5	$e^2 t_2^3$	0; 0
$3d^6$	4	$e^3 t_2^3$	$-0.6\Delta_{tet}$; $-6Dq_t$
$3d^7$	3	$e^4 t_2^3$	$-1.2\Delta_{tet}$; $-12Dq_t$
$3d^8$	2	$e^4 t_2^4$	$-0.8\Delta_{tet}$; $-8Dq_t$
$3d^9$	1	$e^4 t_2^5$	$-0.4\Delta_{tet}$; $-4Dq_t$

[a] unpaired electrons.

Transition from high-spin to low-spin configurations

The idea that there are two possible electronic arrangements for the $3d^4$ to $3d^7$ configurations, with the implication of different crystal field strengths, is puzzling at first when we recall the point charge basis of CFT. This we shall return to later, but in the meantime if we accept the idea that different ligands can exert different crystal field effects, then the relationship between high-spin and low-spin behaviour can be explored.

The transition from high- to low-spin configurations can be represented graphically, Fig. 1.4, which shows what happens when the spin pairing energy decreases as CFSE increases for the $3d^6$ configuration in an octahedral environment. Increasing field strength results in increasing stabilisation for all the configurations $3d^1$ to $3d^9$, excepting the weak field configuration $3d^5$. The slope of the energy lines is determined by the CFSE which is always greater for the strong field situation than for the weak field one.

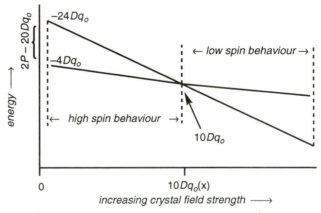

Fig. 1.4 Relationship between pairing energy (P) and CFSE in $3d^6$ octahedral compounds

At some particular value of the crystal field strength, shown in Fig. 1.4 as $10Dq_o(x)$, the energies of the two configurations become equal. For weaker fields, to the left-hand side of the crossover point, the high-spin state is adopted, and to the right-hand side, the low-spin state is more stable. The two states are in equilibrium with each other at $10Dq_o(x)$, and on either side of the *crossover point*, a thermal equilibrium will be established (the difference in energies of the two states is of the order of kT).

Measurement of crystal field splitting

The colour of simple transition complexes is largely determined by d–d electronic transitions which appear in the visible region of the electromagnetic spectrum. To unravel the electronic spectra of transition metal complexes is beyond the scope of this Primer. However, an insight into what happens when a photon of appropriate energy interacts with transition metal d orbitals, and how Δ_{oct} or $10Dq_o$ may be determined, can be obtained from looking at the behaviour of violet $[Ti(H_2O)_6]^{3+}$ in aqueous solution. This ion contains Ti^{III}, with a $3d^1$ configuration. The visible spectrum of this is shown in Fig. 1.5, the broad absorption at 492 nm (20 300 cm^{-1}) being due to the transition $t_{2g}^1 e_g^0 \rightarrow t_{2g}^0 e_g^1$, brought about by the absorption of photons which cause promotion of the unique electron. The energy of this transition just happens in this case to be equivalent to Δ_{oct} or $10Dq_o$, calculated to be 243 kJ mol^{-1}.

By using techniques similar to this, but taking into account electronic configuration, electron–electron repulsion effects, electronic spin, *etc.*, Δ_{oct} or $10Dq_o$ can be worked out for any set of ligands bound to a first-row transition metal ion, and the effect of changing the metal ion while retaining the same coordination environment can also be determined.

Factors influencing the size of the crystal field splitting

We have described how the size of the crystal field is dependent on the *number of ligands* and shown that the splitting in an octahedral field is somewhat more than twice as strong as for a tetrahedral field for the same metal ion and the same ligands ($\Delta_{tet} = 4/9\Delta_{oct}$).

The *oxidation state* of the metal ion is also important. Bearing in mind that the model being used is an electrostatic one, Δ_{oct} increases with metal oxidation state in the order $M^{II} \ll M^{III} < M^{IV}$. This is broadly consistent with the effect of the increased charge on the metal ion drawing the ligands closer, thereby causing a greater splitting of the metal $3d$ orbitals.

Different ligands on the same metal ion cause different degrees of splitting, as can be seen from the data summarised in Table 1.6. There is a steady progression in size of Δ_{oct} in the order Cl$^-$ < H$_2$O < NH$_3$ < CN$^-$. It is possible to list ligands in order of increasing field strength in a *spectrochemical series*:

$$I^- < Br^- < S^{2-} < SCN^- < Cl^- < NO_3^- < F^- < O^{2-} \approx OH^- \approx H_2O < NCS^- <$$
$$NH_3 < en \approx py < bipy < phen < NO_2^- < PR_3 < C_2H_4 \approx CN^- \approx CO$$

Those ligands at the left-hand side of this order (with low Δ_{oct}) are the so-called *weak field ligands*, and those at the opposite end the *strong field ligands*.

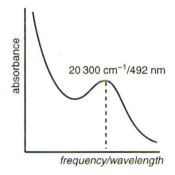

20 300 cm^{-1}/492 nm

frequency/wavelength

Fig.1.5 The visible spectrum of $[Ti(H_2O)_6]^{3+}$ in aqueous solution.

The energy of Δ_{oct} for the absorption at 20 300 cm^{-1} is

$$20\,300 \text{ cm}^{-1} \times \frac{1 \text{ kJ mol}^{-1}}{83.6 \text{ cm}^{-1}}$$

$$= 243 \text{ kJ mol}^{-1}$$

Table 1.6 Value of Δ_{oct} for some octahedral complexes, in kJ mol^{-1}

Ion	$3d^n$	Electronic configuration	6Cl$^-$	6H$_2$O	6NH$_3$	6CN$^-$
				Ligand set		
Cr^{3+}	3	$t_{2g}^3 e_g^0$	164	208	257	318
Mn^{2+}	5	$t_{2g}^3 e_g^2$	90	102	–	
		$t_{2g}^5 e_g^0$				359
Fe^{3+}	5	$t_{2g}^3 e_g^2$	132	171		
		$t_{2g}^5 e_g^0$				419
Fe^{2+}	6	$t_{2g}^4 e_g^2$	–	124		
		$t_{2g}^6 e_g^0$				393
Co^{3+}	6	$t_{2g}^6 e_g^0$	–	248	407	417
Ni^{2+}	8	$t_{2g}^6 e_g^2$	90	102	129	–

1.3 Magnetic properties

Compounds having no unpaired electrons are called *diamagnetic* and are repelled by magnetic fields. Diamagnetism is a universal atomic and molecular property, and is a very small effect. Compounds which are weakly attracted into a magnetic field are called *paramagnetic*. Paramagnetism arises from the intrinsic magnetic properties associated with an unpaired electron (the spin), and the extent of attraction into the applied field is dependent on the number of unpaired electrons. The effect of paramagnetism is about 100 times greater than that of diamagnetism. Many compounds of first-row transition metals fall into this category. Compounds which are very strongly attracted into or repelled by magnetic fields are called *ferro-* or *ferri-magnetic* and *antiferromagnetic*, respectively.

Susceptibility and magnetic moments

When a compound is placed in a magnetic field, it becomes magnetised. The relationship between that magnetisation, M, and strength of the magnetic field, H, is the magnetic induction, B, and is given by $B = \mu_o(H + M)$ where μ_o is the vacuum permeability. This may be rewritten to include κ, the volume susceptibility, which is the ratio of the induced magnetisation to the applied magnetic field:

$$B/H = \mu_o(1 + M/H) = \mu_o(1 + \kappa).$$

However, a more useful quantity is the susceptibility per gram, χ_g, which is κ/ρ where ρ is the density of the compound. The *molar susceptibility* is related to the gram susceptibility by the relationship $\chi_M = \chi_g M$ where M is the molecular weight of the compound.

Taking into account molar volume V, the molar susceptibility can be written as

$$\chi_M = \frac{\text{total magnetisation per mole}}{H} - \frac{\text{average magnetisation per mole} \times N_A}{H}$$

> The unpaired spins on metal ions in ferro- and ferri-magnetic materials exhibit strongly cooperative magnetic effects. Ferromagnetic compounds can behave like permanent magnets. The unpaired spins in anti-ferromagnetic materials are coupled in an antiparallel fashion, leading to a reduction in overall magnetic moments.

where N_A is Avogadro's number.

The magnetic moment of an individual molecule, μ, is related to χ_M as follows:

$$\chi_M = N_A\mu^2/3kT = N_A^2\mu^2/3RT = C/T$$

Here C is the Curie constant and the equation $\chi = C/T$ is Curie's Law. A plot of χ versus $1/T$ should give a straight line. While values of χ may be reported, it is traditional to report the quantity known as the *effective magnetic moment*, μ_{eff}, where μ_{eff} is proportional to $\sqrt{C}$. The effective magnetic moment is related to χ as shown:

$$\mu_{eff} = (eh/mc^2)(\chi T)^{1/2} = 2.828(\chi T)^{1/2}$$

Curie's Law implies that μ_{eff} varies inversely with the square root of temperature, but if significant variations occur, for example a break in the curve, important information can be obtained not only about magnetic behaviour but about the electronic structure of the compound.

Paramagnetic effects originate from two sources: the *spin angular momentum* and *orbital angular momentum* of the electrons. An electron has an intrinsic magnetic moment – it behaves as a minute bar magnet – and the spin of the electron describes the same effect. These minute bar magnets orient themselves in a magnetic field, so giving rise to 'spin-related' magnetic effects. The orbital angular momentum effects arise from the circulation of the electrons about the nucleus under very special circumstances (see below).

In first-row transition metals, orbital angular momentum is relatively unimportant and, to a first approximation, it can be assumed that the major contribution to the effective magnetic moment is spin angular momentum. This leads to a simple formula for calculating the expected magnetic moment based only on the total number of unpaired spins associated with the metal ion. This is known as the 'spin-only' formula (n = total number of unpaired spins):

$$\mu_{eff} = \{n(n + 2)\}^{1/2}$$

Table 1.7 shows the calculated and range of observed values for 'spin-only' magnetic moments of octahedral and tetrahedral first-row transition metal species. There is a reasonable agreement between calculated and observed moments, but some of the deviations can be explained by taking into account orbital angular momentum.

Orbital angular momentum and spin–orbit coupling

As mentioned above, orbital angular momentum is associated with rotation of an electron about the nucleus. The special condition which is attached to this effect in transition metal compounds is that circulation occurs between at least two degenerate orbitals with the same symmetry, providing that they are unequally occupied by electrons.

In an octahedral field, the d_{xz} and d_{yz} orbitals have the same symmetry and are degenerate: a rotation about the z axis by $90°$ converts one into the other. One electron in, say, d_{xz} can move in a clockwise or anticlockwise direction, and either of these movements is equally possible in the absence of a magnetic field. On application of an external field, this degeneracy is removed: the two directions of movement have different energies. This concept lies at the heart of the orbital angular contribution to magnetic moments, and the effect can either reduce or enhance the value of the observed moment based on 'spin-only' considerations.

Table 1.7 Calculated and observed values of μ_{eff} (BM)

Metal	Electronic configuration	'Spin-only' magnetic moment	Range of moments at room temp.
Octahedral			
Ti^{III}	$t_{2g}^1 e_g^0$	1.73	1.6 – 1.75
V^{III}	$t_{2g}^2 e_g^0$	2.83	2.7 – 2.9
Cr^{III}	$t_{2g}^3 e_g^0$	3.88	3.7 – 3.9
Cr^{II}	$t_{2g}^3 e_g^1$	4.90	4.7 – 4.9
Mn^{II}	$t_{2g}^3 e_g^2$	5.92	5.6 – 6.1
	$t_{2g}^5 e_g^0$	1.73	1.8 – 2.1
Fe^{III}	$t_{2g}^3 e_g^2$	5.92	5.6 – 6.1
	$t_{2g} e_g^0$	1.73	2.0 – 2.5
Fe^{II}	$t_{2g}^4 e_g^2$	4.90	5.1 – 5.7
Co^{II}	$t_{2g}^6 e_g^1$	1.73	1.8
Ni^{II}	$t_{2g}^6 e_g^2$	2.83	2.8 –3.5
Cu^{II}	$t_{2g}^6 e_g^3$	1.73	1.7 – 2.2
Tetrahedral			
Mn^{II}	$e^2 t_2^3$	5.92	5.9 – 6.2
Fe^{II}	$e^3 t_2^3$	4.90	5.3 – 5.5
Co^{II}	$e^4 t_2^3$	3.88	4.2 – 4.8

The d_{xy} and $d_{x^2-y^2}$ orbitals are interconverted by a 45° rotation about the z axis but in an octahedral crystal field they are not degenerate so do not give a contribution to the orbital effect on magnetic moments. It is also clear that while the d_{z^2} and $d_{x^2-y^2}$ orbitals are degenerate in an octahedral field, they are not interconverted by a simple rotation and so play no role in orbital angular momentum effects in magnetism.

These arguments can also be applied to a tetrahedral field, but in reverse to take account of the different splitting of the d orbitals. In general, both spin and orbital motions will contribute to the magnetism displayed by a transition metal compound, and their effects may be additive or in opposition to each other. Furthermore, the spin and orbital magnetic effects may interact with each other, giving rise to *spin–orbit* coupling.

Frequently spin–orbit coupling causes a very small reduction in the observed magnetic moments of metals ions with $3d^1$–$3d^3$ configurations, and a more noticeable increase in moments for those with a $3d^6$–$3d^9$ configuration.

1.4 Covalent bonding

The traditional way to introduce concepts of bonding and electronic structure in transition metal chemistry is to begin with crystal field theory (CFT), which is at heart an electrostatic model. While CFT is very successful in explaining and predicting in a very simple way the role of $3d$ electrons in determining many structural features and other physical and chemical properties of compounds of the $3d$ metals, there seems to be a paradox in that most of the interactions between metal and ligands are regarded as 'repulsive'. CFT appears to ignore the obvious fact that the ligands have lone pairs of electrons which are donated to the central transition metal. Furthermore, the unexpected stability of an important sub-class of transition metal compounds, the metal carbonyls, $[M_n(CO)_m]$, cannot be satisfactorily rationalised by CFT.

In order to advance the understanding of the properties of transition metal compounds, incorporating satisfactory explanations based on the splitting of the $3d$ orbitals in octahedral, tetrahedral or planar geometries but taking into account the ability of ligand and metal orbitals to *overlap* with each other, it is necessary to introduce a covalent model based on simple *molecular orbital* methods.

Metal–ligand σ bonding in octahedral geometry

First-row transition metals use their $4s$, $4p$ and $3d$ orbitals in bonding with ligands, but the way these orbitals combine with the σ-donor orbitals of the ligands, by a linear combination of atomic and ligand orbitals, is beyond the scope of this text. A mathematical treatment, Group Theory, is the most effective way of addressing this matter, since it uses the symmetry of the complex and of the orbital sets to determine which orbitals can overlap to form bonding and antibonding molecular orbitals. Some of Group Theory's symbolism will be used here as a shorthand notation for groups of ligand orbitals and particular molecular orbitals.

The metal $4s$ orbital overlaps with all six of the ligand donor orbitals forming a bonding and antibonding pair of molecular orbitals; these are labelled a_{1g}. The metal $4p$ orbitals will overlap with particular pairs of ligand orbitals, forming a triply degenerate set of molecular orbitals, with a corresponding antibonding set, labelled t_{1u}. Of the $3d$ orbitals, the e_g set will interact with six and four ligand orbitals, respectively, giving the e_g set of molecular orbitals. The bonding interactions are shown in Fig. 1.6. The remaining $3d$ orbitals, which point between the ligands, cannot overlap in a σ-fashion with the ligand orbitals, and so are formally non-bonding in an octahedral species.

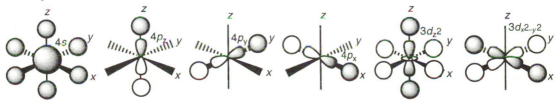

Fig. 1.6 Bonding combinations of ligand σ orbitals with metal orbitals

The molecular orbital diagram for an octahedral complex in which the metal and ligands interact solely by σ bonding is shown in Fig. 1.7. This provides a link between the molecular orbital approach and CFT. The non-bonding t_{2g} and weakly antibonding $e_g{}^*$ sets of orbitals correspond to the t_{2g} and e_g levels in the crystal field splitting diagrams for an octahedral species, and the crystal field splitting energy is directly comparable.

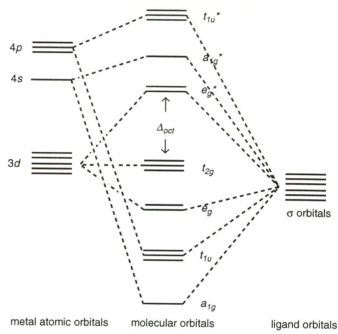

Fig. 1.7 Energy level diagram for an octahedral metal complex, showing only σ–bonding interactions

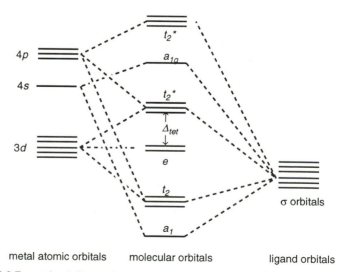

Fig. 1.8 Energy level diagram for a tetrahedral metal complex, showing only σ–bonding interactions

Metal–ligand σ interactions in tetrahedral geometry

A similar energy level diagram may be constructed for a tetrahedral species, Fig. 1.8. In this geometry, four orbitals comprise the ligand group: one of these overlapping with the $4s$ orbital, and the remainder interacting with the $4p$. The t_2 orbitals also interact with the ligand orbital set, in contrast to the situation in the octahedron, and the e orbitals are non-bonding.

Once again, the relationship between CFT and the inclusion of covalency can be seen to be broadly equivalent. The splitting between the e and t_2 levels in both models, Δ_{tet}, is equivalent.

Effect of π bonding in octahedral complexes

It can usually be assumed that each local metal–ligand interaction will include a contribution from π bonding: relatively simple ligands can act as π donor or π acceptors to a varying degree depending on the nature of the ligand. The π orbitals on these ligands are generally derived from p orbitals on the ligating atom, although involvement of d orbitals in P- and S-donor atom ligands is also possible.

In an octahedral system, there are two p orbitals associated with each ligand. If the ligand is situated on the z axis, then the orbitals capable of π bonding are p_x and p_y. If the ligand is located on the x axis, the p orbitals involved are p_z and p_y (Fig. 1.9). An inspection of the shapes of the d orbitals and these potentially π-bonding p orbitals will show that so far as the metal is concerned, only one set of p orbitals is important in terms of π interaction with the metal $3d$ orbitals. The type of interaction, bonding and antibonding, is shown in Fig. 1.10. There are three such sets of interactions, involving the d_{xz}, d_{yz} and d_{xy} orbitals.

In constructing an energy level diagram to include potential π bonding arising from the ligands, we can simplify the overall scheme by leaving out those contributions from groups of ligand $p(\pi)$ orbitals which are relatively uninvolved in bonding with the metal $3d$ orbitals. In this simplified energy level diagram (Fig. 1.11) it is necessary to take into account whether the ligand π orbitals are filled or empty, since this makes an enormous difference to the relative energies of the resulting molecular orbitals.

Figure 1.11(a) shows the situation where the ligand π orbitals are filled and relatively low-lying with respect to the metal $3d$ orbital energies. The bonding e_g and t_{2g} levels are likely to be filled because they are largely constituted by the σ lone pairs and electrons in the π orbitals of the ligands. The crystal field splitting, Δ_{oct}, is the energy gap between the formally antibonding t_{2g}^* and e_g^*. The effect of π bonding is to decrease Δ_{oct} relative to the situation where there is only σ bonding, mainly because π bonding increases the energy of the t_{2g}^* levels (which are mainly metal-based).

Figure 1.11(b) depicts what happens when the ligand π orbitals are empty and lying significantly higher in energy than the metal $3d$ orbitals. The lowest (e_g) molecular orbital will be filled because it is principally made up of ligand lone pairs. The splitting parameter Δ_{oct} is now the energy gap between the formally bonding t_{2g} and antibonding e_g^* levels, and its magnitude is significantly greater than that when the ligand π orbitals are filled or, indeed, when only σ bonding occurs. This is caused by the very

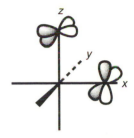

Fig. 1.9 $p\,\pi$ orbitals, showing only one pair on the z and one pair on the x axes: all ligands have such orbitals. The six ligands are capable of providing 18 orbitals for potential bonding in an octahedral complex: six σ orbitals and 12 $p\pi$ orbitals. Of these only the σ and six $p\pi$ orbitals are significant in their interaction with the metal d orbitals

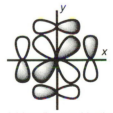

(a) bonding combination

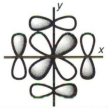

(b) antibonding combination

Fig. 1.10 (a) Bonding and (b) antibonding π interactions between a $3d_{xy}$ and p_x and p_y orbitals

strong interaction between the ligand $t_{2g}(\pi)$ orbitals and the metal d_{xz}, d_{yz}, and d_{xy} orbitals which leads to an increased stabilisation of the formally bonding t_{2g} molecular orbitals.

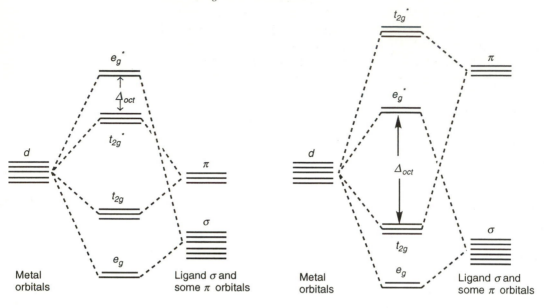

(a) Interaction between 3*d* orbitals and the filled σ and filled t_{2g} (π) ligand orbitals

(b) Interaction between 3*d* orbitals and filled σ and empty t_{2g} (π) ligand orbitals

Fig. 1.11 Effect of inclusion of π bonding on *d* orbital splitting

The energy level diagrams shown in Figs 1.8 and 1.11 provide an improved explanation for high-spin/low-spin behaviour and the spectrochemical series. That series places the good *π-donor* ligands at the left-hand side in the order of increasing size of Δ_{oct} where high-spin behaviour predominates, and the effective *π acceptor*-ligands at the right-hand side where Δ_{oct} is substantial and low-spin behaviour is observed.

Some structural consequences of electronic configuration

As described in Section 1.2, it is possible to derive the 3*d* orbital splitting diagram for a tetragonal and, thence, a planar geometry by imagining the effect of withdrawing ligands along the *z* axis of a hypothetical octahedron. This provides a logical way of working out what will happen to the *d* levels on distortion, but tells us nothing about when to expect such a distortion. The clue to this lies in the population of the 3*d* levels.

For a $3d^9$ configuration in a strictly octahedral environment, the e_g level is degenerate and contains three electrons. This situation is inherently unstable and a distortion of the environment occurs to remove that degeneracy. This is the *Jahn–Teller effect*. The distortion leads either to an elongation or a shortening of the metal–ligand bond lengths with respect to the pure octahedral arrangement expected. The tetragonal distortion effect shown in Fig.1.3 is due to elongation, which places the $3d_{x^2-y^2}$ orbital above the $3d_{z^2}$ orbital The effect of compression (bond-shortening along the *z* axis) would reverse this order, and also have a commensurate effect on the former t_{2g} levels, giving the splitting arrangement shown in Fig. 1.12. It is not possible to predict reliably when elongation or compression will occur.

Apart from the regular occurrence of Jahn–Teller distortion in Cu^{II} complexes because of the metal's $3d^9$ configuration, similar distortions can be found in octahedral Cr^{II} ($t_{2g}^3 e_g^1$) and, in principle, in low-spin Co^{II} ($t_{2g}^6 e_g^1$). Jahn–Teller effects arising from incomplete t_{2g} orbitals are relatively insignificant because the ligands are not directly in contact with these orbitals.

In tetrahedral species, orbital degeneracy in the t_2 levels would be expected to produce small distortions, particularly for the d^3, d^4, d^8 and d^9 configurations. Tetrahedral $3d^3$ and $3d^4$ complexes are very rare and while tetrahedral $3d^8$ complexes (*e.g.*, $[NiX_4]^{2-}$) are more common, they do not appear to be especially distorted. However, tetrahedral Cu^{II} complexes are significantly distorted, frequently flattening towards planar geometries.

In principle, the number of unpaired electrons in a tetragonally distorted species could be different to that of its hypothetical octahedral precursor, but this is virtually impossible to test. However, the ultimate in tetragonal distortion is the removal of both ligands from the z axis, and formation of a planar species. For the $3d^8$ configuration, there will be no unpaired electrons: $(3d_{xz}, 3d_{yx})^4 (3d_{xy})^2 (3d_{z^2})^2 (3d_{x^2-y^2})^0$. This provides an explanation in nickel(II) chemistry for the paramagnetism of octahedral and tetrahedral complexes and the diamagnetism of planar species (Fig. 1.13).

1.5 Colour and electronic spectra

This is a topic whose detailed explanation is beyond the scope of this Primer. Nevertheless some comments are necessary on what is, after all, one of the main distinguishing features of d-block elements: that many of their compounds are coloured. To provide an adequate explanation of the occurrence of one absorption in the visible range of all but the simplest d-orbital configuration, *viz.*, d^1, requires the use of Group Theory and also a deeper knowledge of the way in which spin and orbital angular momentum interact. Accordingly, only some very simple discussion is provided.

In general the promotion of an electron from one orbital to another or, more exactly, the excitation of a molecule from its ground state to an electronic excited state, corresponds to the absorption of light in the near-infrared (NIR), visible (VIS) or ultra-violet (UV) region of the electromagnetic spectrum. In transition metal compounds the absorption bands in the NIR and VIS regions are generally weak and associated with transitions localised on the metal atom, *i.e.*, the d–d transitions (these are approximately 100 to 1000 times weaker than absorptions in conventional dyestuffs). The UV transitions, in contrast, are strong (intense) and are due to electronic transitions between atoms, *i.e.*, within the ligands or between ligands (n–π^* or π^*–π^* transitions) or between metal and ligand (metal-to-ligand or ligand-to-metal transitions). These absorptions are called *charge transfer* bands.

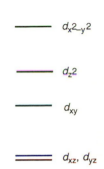

Fig. 1.12 Splitting of $3d$ orbitals in a compressed octahedral field (short metal–ligand bond lengths along the z axis)

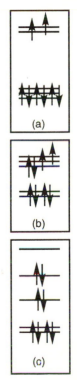

Fig. 1.13 $3d$ orbital configurations for Ni^{II} for (a) octahedral, (b) tetrahedral, and (c) planar geometries

d–d transitions

The *d–d* transitions are weak relative to charge transfer transitions. They are in fact formally forbidden, although obviously not totally so! There are two selection rules which broadly govern the observance or otherwise of *d–d* transitions. The first of these, the *Laporte rule*, relates to the symmetry of the orbitals involved in the electronic transition and of the compound involved. For an electronic transition to occur, the electron must move either from a *gerade* orbital to an *ungerade* orbital, or *vice versa*. In an octahedral environment, which has a centre of symmetry, the $3d$ orbitals are labelled *gerade*, *g*, i.e., t_{2g} and e_g, and so *d–d* transitions are not allowed. A tetrahedral environment, however, does not possess a centre of symmetry, and so this selection rule does not apply, and *d–d* transitions are not forbidden.

The second rule, known as the *spin selection rule*, requires that excitation or promotion of an electron from one orbital to another can only occur if the orientation of the electron spin is conserved. In other words this means that the electron spin must be the same in the ground and the excited state: electron spin cannot invert in the process of excitation.

So how can one explain the observation of electronic transitions in the visible region? This is done simply by remembering that the molecules under investigation are not totally rigid: they undergo normal vibrations which destroy the ideal octahedral symmetry of the compound. This means that the compound is not centrosymmetric at all times, and so the Laporte selection rule does not apply rigidly.

Some general points can be made about the electronic spectra of first-row transition metal compounds.

(a) The spectra of most octahedral complexes of ions having the configurations $3d^1$, $3d^4$, $3d^6$ and $3d^9$ are characterised by a single absorption, while those for many complexes with $3d^2$, $3d^3$, $3d^7$ and $3d^8$ configurations can have as many as three principal absorptions.

(b) The spectra of high-spin $3d^5$ complexes (*e.g.*, high-spin Mn^{2+} and Fe^{3+}) are very weak, although they contain a large number of absorptions. The reason is inherent in the spin selection rule which forbids transitions to orbitals with simultaneous inversion of spin, and the electronic configuration requires single occupancy of all $3d$ orbitals, *viz.*, $t_{2g}^3 e_g^2$.

(c) For those metal ions with $3d^0$ and $3d^{10}$ configurations, no *d–d* transitions will be observed. However, $[MnO_4]^-$ is deep purple in solution despite the fact that the ion contains Mn^{VII} which has a $3d^0$ configuration. The effect is due to charge transfer.

Charge transfer transitions

An electronic transition which results in the displacement of charge from one atom to another is a charge transfer process. These are broadly of three types: intraligand transitions, ligand-to-metal charge transfer, and metal-to-ligand charge transfer.

> Orbitals which have a centre of symmetry when in a molecule which is also centrosymmetric are *gerade* (*g*) while those which do not are labelled *ungerade* (*u*).

Intraligand transitions

These are associated with the $n-\pi^*$ and $\pi^*-\pi^*$ transitions within ligands around a transition metal. They are almost invariably in the UV region.

Ligand-to-metal charge transfer transitions, LMCT

This is a transition in which a ligand electron is transferred to a metal orbital. It can sometimes be difficult to ascertain which ligand orbital is involved in the LMCT process, and indeed, some are formally non-bonding with respect to the $3d$ orbitals. These transitions are fully allowed (they must be 'g' to 'u' or 'u' to 'g' in terms of the Laporte rule) and consequently the absorptions are very strong. LMCT processes may occur in complex ions, such as $[MnO_4]^-$ (violet), $[CrO_4]^{2-}$ (deep yellow, Cr^{VI}, $3d^0$) and $[Cr_2O_7]^{2-}$ (deep orange, Cr^{VI}, $3d^0$). They may also occur in more 'ionic' solids such as the inorganic pigments chrome yellow, $PbCrO_4$, and red and yellow ochres, Fe^{III} oxides, where oxygen p (with 'u' symmetry) to metal d (with 'g' symmetry) LMCT transitions are responsible for the intense colours of the compounds. Similar explanations account for the green colour of copper (II) chlorides and the near-black colour of the comparable bromides.

Metal-to-ligand charge transfer transitions, MLCT

In this type of transition, an electron is transferred from the metal to an orbital largely based on a ligand. It is, in a sense, the opposite of LMCT, and it can sometimes be difficult to differentiate between the two charge transfer processes. MLCT is encountered in compounds containing π acceptor ligands such as CO, CN^- and bipy.

2 Metals and solid compounds

2.1 Occurrence of the 3*d* elements in nature

While the *transition elements* as a whole are widely distributed throughout the Earth's crust, by far the most abundant is iron. In fact, of all the elements encountered within the crust, iron is fourth after oxygen, silicon and aluminium, comprising about 6% of the total crustal composition. This is not so surprising given the widely accepted belief that the Earth's metallic core is mostly made up of this element (see below).

The next most abundant *first-row transition elements* are titanium (*ca.* 0.6%), and manganese (*ca.* 0.1%) while the remainder (V, Cr, Co, Ni and Cu) are more sparsely distributed (less than 0.01%).

The majority of the elements occur in minerals as oxides, hydrated oxides and carbonates, although sulphides and arsenides are also significant. The principal minerals and other sources of the elements are shown in Table 2.1.

Table 2.1 Occurrence of first-row transition elements

Metal	Mineral sources
Ti	*Rutile* TiO_2, *ilmenite* $FeTiO_3$
V	*Patronite* (complex sulphide); *vanbadinite* $Pb_5(VO_4)_3Cl$; *carnotite* $K(UO_2)VO_4.3/2H_2O$
Cr	*Chromite* $FeCr_2O_4$; *crocoite* $PbCrO_4$; *chrome ochre* Cr_2O_3
Mn	As the silicate; *pyrolusite* MnO_2; *hausmannite* Mn_3O_4; *rhodochrosite* $MnCO_3$
Fe	*Haematite*, Fe_2O_3; *magnetite* Fe_3O_4: *limonite* $FeO(OH)$; *siderite* $FeCO_3$; *pyrites* FeS_2
Co	*Smaltite* $CoAs_2$; *cobaltite* $CoAsS$; *linnaeite* Co_3S_4; in association with Ni in arsenic-containing ores
Ni	*Laterites* such as *garnierite* $(Ni,Mg)_6Si_4O_{10}(OH)_8$ and *nickeliferous limonite* $(Fe,Ni)O(OH).nH_2O$; *sulphides* such as *pentlandite* $(Ni,Fe)_9S_8$; associated with Cu, Co and other precious metals in combination with As, Sb and S
Cu	*Chalcopyrite* $CuFeS_2$; *copper glance* or *chalcotite* Cu_2S; *cuprite* Cu_2O; *malachite* $Cu_2CO_3(OH)_2$

Vanadium is found naturally in some living systems, particularly sea squirts, tunicates and some mushrooms, but its role is far from clear. The element also occurs in certain petroleum deposits, particularly from Venezuela, and can be extracted from them as a vanadyl porphyrin compound.

Trace amounts of chromium, as oxides, are responsible for the characteristic colours of ruby and emerald. Gem-quality corundum (γ-Al_2O_3) containing traces of iron is commonly known as sapphire. Manganese occurs in over 300 different and widely distributed minerals, and as a result of weathering, colloidal particles of manganese and other metal (iron, copper, cobalt) oxides are continually washed into the sea. On agglomeration and being compacted on a geological time scale, 'manganese nodules' have been formed which, when dried, contain between 15 and 30% manganese. This is on the verge of being economically viable for extraction.

Cobalt and nickel frequently coexist in minerals, often in association with copper and lead, and in combination with arsenic, antimony and sulphur. Elemental nickel is also found, alloyed with iron, in meteors and is believed to occur in the Earth's core.

2.2 Structure of solids

The structures of many solids can be conveniently represented in terms of the close-packing of spheres, the spheres representing, to a first approximation, atoms, simple or complex ions, or even molecules.

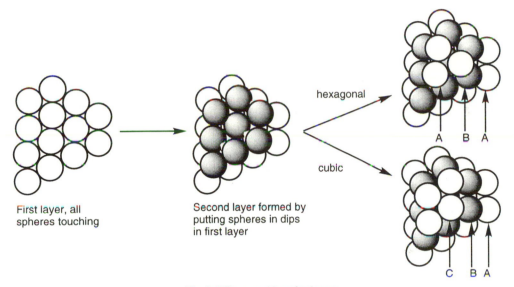

Fig. 2.1 Close-packing of spheres

Metals are the easiest solids to describe in this way, since the atoms are all of the same size and they can pack closely together forming two distinct arrangements: *hexagonally close-packed* (hcp) and *cubic close-packed* (ccp). In the latter the spheres have a face-centred cubic (fcc) arrangement (see Fig.2.4, p.25). The former is shown in Fig. 2.1 where the third layer is superimposed directly over the first, giving an ABA... pattern of layer alternation. In the latter, the third layer is imposed above the gaps in the first layer, giving an ABC... pattern of layer alternation.

There are two types of 'holes' in these close-packed structures: the tetrahedral (Fig. 2.2(a)) and the octahedral 'hole' (Fig. 2.2(b)). The former is generated by a planar triangle of touching spheres which are capped by the fourth sphere lying in the dip between them. If there are n spheres in the solid, there are $2n$ tetrahedral holes (the apex of the 'tetrahedron' can be 'up' or 'down'). The octahedral 'holes' are generated by two oppositely directed triangles of touching spheres and if there are n spheres, there are n octahedral 'holes'.

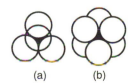

(a)   (b)

Fig. 2.2 Holes in close-packed structures

Fig. 2.3 Body-centred cubic structure

Although the hexagonal and cubic close-packed arrangements are frequently met, there are other ways in which spheres can pack almost as efficiently, body-centred cubic (bcc) being relatively common in metals (Fig. 2.3). Metals which are tightly close-packed at low temperatures adopt a body-centred structure at high temperatures. This is due mainly to the requirement for a less-compact structure as atomic vibrations increase with temperature.

Alloys are formed by melting two or more metals and then cooling the mixture. If the mixture results in a random distribution of the atoms of one metal among those of the others, then a solid solution is formed. There are two types of such solutions: substitutional and interstitial. The former is produced when the atoms of the component metals have very similar radii and the latter when the atoms of one component are very small relative to the other's. The former situation occurs in alloys of nickel and copper, and the latter occurs with non-metals such as H, B, C or N, these atoms being accommodated in the 'holes' in the host metal whose basic structure is preserved.

When molten metal mixtures are cooled, phases with definite structure can be formed which may not be related to the structures of the alloy components. These materials are intermetallic compounds, and β-brass (CuZn) falls in this category (see also p.25).

2.3 Metals: preparation and physical properties

Titanium

The metal cannot be made by reduction of TiO_2 with carbon due to the formation of the extremely stable carbide TiC and, furthermore, hydrogen reduction of the dioxide at 900°C gives only Ti_2O_3. However the metal is obtained by the *Kroll process* which involves chlorination of iron titanate in the presence of carbon, to give $TiCl_4$.

$$2FeTiO_3 + 7Cl_2 + 6C \rightarrow 2TiCl_4 + 2FeCl_3 + 6CO$$

The reaction is carried out at 900°C, and the $TiCl_4$ is fractionally distilled from $FeCl_3$ and other impurities. The tetrachloride is then reduced to titanium metal by magnesium at 1000°C, the magnesium chloride formed being volatilised out, and excess magnesium dissolved away in dilute HCl.

$$TiCl_4 + 2Mg \rightarrow Ti + 2MgCl_2$$

The titanium, formed as a sponge, can be cast into ingots, but very pure metal is best obtained by decomposition of vaporised TiI_4 by a hot wire.

The metal is less dense than iron, stronger than aluminium, and almost as corrosion resistant as platinum. For these, and other reasons, the metal is particularly useful in the construction of very high performance combustion engines, aircraft frames, marine equipment and industrial plants.

At normal temperatures, titanium is hexagonal, but at about 820°C it undergoes a phase change to the β form which has a cubic structure.

Vanadium

Vanadium ores are converted to V_2O_5 or metavanadates ($[VO_3]^-$) by roasting with carbon. Since the metal is extremely reactive towards carbon, nitrogen and oxygen, particularly at high temperatures, it is difficult to produce free of impurities. The pure metal can be made by reduction of VCl_4 or V_2O_5, by the thermal decomposition of VI_4, or by electrochemical reduction.

Vanadium, like titanium, is corrosion resistant and hard. Its main use is in alloying with steels giving ductile and shock-resistant *ferrovanadium*.

> At normal temperatures, vanadium has a body-centred cubic lattice.

Chromium

Reduction of chromite, $FeCr_2O_4$, by carbon in an electric furnace gives the useful alloy *ferrochromium*:

$$FeCr_2O_4 + 4C \rightarrow 2Cr + Fe + 4CO$$

If pure chromium is needed, chromite is converted into chromate ($[CrO_4]^{2-}$) by alkali and oxygen. This is transformed into dichromate which is reduced by carbon to Cr_2O_3, the pure metal being obtained by reduction of the oxide with Al or Si.

The metal is hard and brittle but has low ductility at ordinary temperatures which significantly reduces its use when pure. However, chromium is extremely resistant to normal corrosive agents which is why it has been extensively used as an electroplated protective coating. Most chromium metal is used in the production of non-ferrous alloys.

> Chromium has a body-centred cubic structure.

Manganese

The metal is obtained by reduction of various oxides by Al, and its physical and chemical properties are quite similar to iron although it is harder and more brittle. Its main use is in alloying with iron to give *ferromanganese*, important in the hardening of steels.

> Manganese has four allotropes, but only the α form is stable at room temperature, having a body-centred cubic structure.

Iron

Iron is made industrially by reduction with coke in a blast furnace. The reducing agent is CO which converts the iron oxides to the metal and is itself oxidised to CO_2. Limestone is added to the process to remove sand and clays as slag. The molten iron produced is known as 'cast-iron' or 'pig-iron' and is impure, containing about 4% carbon and variable amounts of silicon, manganese, phosphorus and sulphur. Synthesis gas ($H_2 + CO$) may also be used as a reducing agent, at a lower temperature than the coke process, the metal produced having sponge-like characteristics. This technology is an alternative when coke is not easily available, and is appropriate for small-scale production of steel *via* electric arc furnace refining (see below).

The 'cast-iron' is hard and brittle, but if the non-metallic impurities are removed by oxidation using Fe_2O_3, 'wrought-iron' can be produced which is tougher and more malleable, and ideal for mechanical working. This is rarely carried out now, since the bulk of 'cast-iron' is converted to steel by the basic oxygen process (BOP) or by open hearth or electric arc furnace techniques. These essentially employ high temperature oxidation *via* O_2 which drives off most of the non-metals, leaving between 0.5 and 1.5% carbon and traces of

> Iron exists as several allotropes: a room temperature form, α, which is body-centred cubic, and higher temperature forms (β, apparently body-centred cubic; γ, face-centred cubic; and δ, body-centred cubic).

sulphur and phosphorus, without significant oxidation of the iron and trace metals. This affords 'mild steel', which is stronger and more workable than 'cast-iron', but which can be hardened by alternately heating to red heat and quenching in water or mineral oil, and 'tempered' by reheating to 200-300°C and cooling more slowly. The hardness, resilience and ductility of the steel can be controlled by varying the temperature and rate of cooling, and through the precise composition, particularly the carbon content, of the steel. Alloying with other metals also dramatically affects the properties of steel.

On a small scale, iron can be produced by the reduction of the pure oxide or hydroxide with hydrogen, or by reacting the impure metal with CO under pressure, which gives $[Fe(CO)_5]$ (Section 5.1). This carbonyl may subsequently be thermally decomposed to the pure element. It is also produced by electro-deposition from aqueous solutions of Fe^{2+} salts.

The pure metal is not especially hard, and is readily attacked by moist air forming hydrated oxides, *i.e.*, rust. It is easily magnetised. When finely divided, iron is pyrophoric and reacts with ammonia at 400°C giving a nitride which is the basis of 'nitriding' – another process for the hardening of steels.

Cobalt

Cobalt is frequently obtained as a by-product of copper production. However, when arsenic-containing ores are roasted (volatile As_2O_3 being recovered as a valuable side-product), the mixed ores can be leached with sulphuric acid which dissolves out the iron, nickel and cobalt, leaving copper behind. The iron and nickel salts are precipitated from the acid solution by addition of lime, and the cobalt is separated by precipitation with NaOCl, which affords basic cobalt oxide. On heating, this breaks down into Co_2O_3 and CoO which afford the metal on reduction with red-hot charcoal.

Pure cobalt is harder and less reactive than iron. It may be combined with other metals to give magnetic alloys, *e.g.*, *alnico* (Al, Ni and Co) which is a light, permanent magnet up to 25 times more powerful than magnetised iron. Addition of cobalt to iron affords a high-temperature alloy.

> Cobalt exists as two allotropes: α which has a hexagonal close-packed structure, and β which is face-centred cubic.

Nickel

The Earth's core is believed to be made up of an iron–nickel alloy, and elemental nickel is found alloyed with iron in meteorites. The industrial production of nickel is complicated, depending on the source of the ore, which may be oxide or sulphide. In general, the sulphides are roasted in air to give NiO which is then reduced by carbon to impure metal. Refinement can be achieved either electrolytically or by reaction with CO at about 50°C and normal pressures, which affords $[Ni(CO)_4]$ (Section 5.1). The tetracarbonyl is easily thermally decomposed to extremely pure metal.

The pure metal is malleable and ductile, and has relatively high thermal and electrical conductivities. It is used in the production of both ferrous and non-ferrous alloys, the former including steels used in armoured plating and stainless steels (up to 8% Ni). The latter alloys are extensive, including *EPNS* or electroplated nickel silver which is used in tableware, corrosion-resistant *Monel metal* (58% Ni, 32% Cu, traces of Mn and Fe), *nichrome* (60% Ni, 40% Cr) which has a very small temperature coefficient of

> Nickel has a face-centred cubic structure.

electrical resistance, and *invar* (Ni, Fe) which has a very small coefficient of expansion.

Nickel is quite resistant to corrosion and is sometimes used as an undercoat for electroplated chromium. It reacts extremely slowly with F_2 and with corrosive fluorides and so is used in equipment for handling these reactive substances.

Copper

Crushed copper-containing ores are concentrated by froth-flotation, the main impurities being Fe, Ni, Au and Ag. The concentrate is then melted with sand (SiO_2) at 1400–1450°C, iron silicate slags and a matte containing Cu_2S and FeS being formed. The matte is further heated with sand and oxygen which removes most of the iron as oxide in slag, leaving Cu_2O which is then reduced to blister copper. The metal is purified electrolytically. It may also be extracted from aqueous acid solution using organic solvents containing chelating ligands.

The metal is non-magnetic, strong but soft, malleable, ductile, fairly corrosion resistant and has high thermal and electrical conductivity. The last two properties are mainly responsible for the enormous usage of the metal in electric motors, electronic equipment, plumbing installations and building construction, and in household appliances, jewellery and coinage.

> Copper has a face-centred cubic structure.

Copper forms many alloys, the most common of which are *bronze* (Cu + 7–10% Sn), *brass* (Cu + 10–40% Zn which causes the colour to change from a bronze hue to reddish yellow), *nickel silver*, *Monel metal*, aluminium bronzes (Cu + Al), phosphor bronzes (Cu + up to 10% Sn and 0.35% P) and *cupro-nickels* (up to 80% Cu, the rest mainly Ni) used in 'silver' coinage. Brass is also used extensively because it is stronger and more corrosion-resistant than pure copper.

In moist air, copper metal assumes a green coating or patina of hydroxo-carbonates and hydroxo-sulphates, and many old buildings whose roofs were originally constructed of the metal now have this malachite-green colour as a result of atmospheric corrosion.

2.4 Binary and ternary compounds

Structures of M_nX_m

Sodium chloride or rock salt structure, NaCl

NaCl crystallises in a fcc structure (Fig. 2.4), the coordination geometry of the metal and anion being six. Among the first-row transition metal compounds which adopt this structure are the *monoxides*, MO (M = Ti, Fe and Ni). Pyrites, FeS_2, can be described as a distorted NaCl-type structure in which rod-shaped S–S^{2-} ions are centred on the Cl positions, oriented so that they are inclined away from the cubic axis. A similar structure is adopted by MS_2 (M = Mn, Co and Ni).

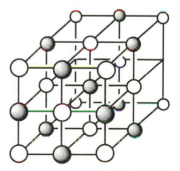

Fig. 2.4 Face-centred cubic structure

Caesium chloride structure, CsCl

CsCl crystallises in a cubic arrangement in which the coordination number of the metal and ion is each eight. It is much less common than the sodium chloride type, and numerous alloys have this structure (*e.g.*, CuZn, AlFe).

Zinc blende or sphalerite and wurtzite structures, ZnS

In a close-packed lattice of sulphide ions, zinc ions usually occupy tetrahedral holes because they are smaller than sulphide ions. If the latter form a ccp array, then the zinc blende structure is formed, but if they form a hcp array then the wurtzite form is adopted, each ion being four-coordinated in both structures. CuCl adopts the zinc blende structure whereas MnS has a wurtzite structure.

Nickel arsenide structure, NiAs

This structure is based on an expanded, distorted hcp anion array, the cations occupying octahedral 'holes'. The arsenic ions are trigonal prismatically coordinated by the nickel ions. Many binary metal *monosulphides* MS (M = Ti, V, Cr, Fe, Co, Ni) adopt this structure, as well as many examples of MSe, MTe, MP, MAs, and MSb.

Fluorite structure, CaF$_2$

This structure is based on an expanded fcc lattice of the cations. This generates two types of tetrahedral 'holes', both of which are occupied by the anions. The cation adopts eight-coordination with a cubic arrangement of F$^-$ around the Ca^{2+}. The anion is four-coordinate, but if the positions and numbers of cations and anions are reversed, the *antifluorite* structure is formed.

Rutile structure, TiO$_2$

The structure is based on an hcp array of oxide ions, the cations occupying only half the available octahedral 'holes'. This results in a tetragonal structure in which each oxide ion is shared by three TiO$_6$ octahedra. This structure is adopted by some *dioxides* MO$_2$ (M = Cr, Mn) and *difluorides* MF$_2$ (M = Cr, Fe, Co, Ni). CrF$_2$ (Cr^{2+}, $t_{2g}^3 e_g^1$) has a distorted structure in which 4F$^-$ are closer to the metal than the other two, a consequence of Jahn–Teller distortion (p.16).

Cadmium dihalides, CdX$_2$ (X = Cl or I)

The dichloride is similar to the ccp array of chloride ions in NaCl, but with every other octahedral 'hole' filled by Cd^{2+}. The di-iodide has hcp I$^-$ ions, also with Cd^{2+} in half of the accessible octahedral 'holes'. The effect of the half-occupancy of all the available octahedral 'holes' in these lattices is to generate a layer structure (Fig. 2.5). The *dichlorides* MCl$_2$, M = Mn–Ni, have the CdCl$_2$ structure while many *dibromides* and *di-iodides* (Ti–Co, NiBr$_2$, CuBr$_2$) and *hydroxides* M(OH)$_2$ (M = Mn–Ni) adopt the CdI$_2$ structure.

Rutile is an extremely important white pigment used in paint, surface coatings of paper, and as a filler in rubbers and plastics. It has a very high refractive index and chemically is relatively inert. *Manganese dioxide* is used in the manufacture of dry-cell batteries, in the colourisation of bricks and in the oxidation of aniline to hydroquinone, important in the production of dyes and paints.

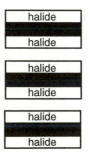

Fig. 2.5 Layer structures in CdX$_2$

Bismuth tri-iodide, BiI₃

Bismuth tri-iodide, BiI$_3$

This is also a layer structure, having an hcp array of iodide ions with two-thirds of the octahedral 'holes' in each alternative pair of layers occupied by cations. Many *trihalides* have this structure, *e.g.*, $TiCl_3$, VCl_3, $CrBr_3$ and $FeCl_3$.

Corundum, α-Al$_2$O$_3$

The α-form of Al_2O_3 has an hcp array of oxide ions with two-thirds of the available octahedral 'holes' filled by cations. Many oxides of the type M_2O_3 (M = T, V, Cr, Fe) have a corundum structure.

Gemstones are formed when Al_2O_3 crystals contain transition metal impurities. For example, ruby (red) contains Cr^{III}, oriental amethyst (violet) contains Cr^{III}/Ti^{IV}, and sapphire (blue) contains $Fe^{II/III}$ and/or Ti^{IV}.

Structures of $M_nX_mY_p$

Perovskites, ABO$_3$

In the mineral perovskite, $CaTiO_3$, the A cation (Ca^{2+}) is much larger than the B cation (Ti^{4+}), and the oxide ions and calcium form a ccp array with the small Ti^{4+} ion occupying those octahedral 'holes' formed exclusively by oxide ions. The structure is frequently distorted, partly as a consequence of the small size of the B cation, and is adopted by many ABO_3 oxides. The charges on the cations can vary so long as they total six, and typical examples include $BaTiO_3$, and the mixed fluoride $KNiF_3$.

Ilmenite, FeTiO$_3$

This is closely related to the corundum structure except that the cations are of two different types. ABO_3 oxides adopt this structure when the two cations are of approximately the same size, but again, the total charge on A and B must total six. $MgTiO_3$ and $CoTiO_3$ are similar to $FeTiO_3$.

Spinels, AB$_2$O$_4$

The mineral spinel is $MgAl_2O_4$ which has a structure based on a ccp array of oxide ions in which the metals are dispersed in tetrahedral and octahedral 'holes'. There are twice as many tetrahedral as there are octahedral 'holes' and both are large enough to accommodate first-row transition and other metal cations. The octahedral 'holes' are larger than the tetrahedral 'holes', so larger metal ions would normally occupy them in the lattice, but there are exceptions.

In natural spinel, one-eighth of the tetrahedral 'holes' are occupied by Mg^{2+} ions and one-half of the octahedral 'holes' are occupied by Al^{3+}. This structure is adopted by most mixed metal oxides of the type $M^{II}M^{III}_2O_4$ (*e.g.*, $FeCr_2O_4$, Co_3O_4) and by some of the type $M^{IV}M^{II}_2O_4$ (*e.g.*, $SnCo_2O_4$). This structural type is sometimes represented as $A[B_2]O_4$ where the square brackets enclose the ions in the octahedral 'holes'. This is the *normal spinel* structure.

If a crystal is introduced into an electric field and it experiences a reversible contraction or elongation, or the crystal generates an electric field when it is stressed (pressure), it is exhibiting the *piezoelectric effect*. This behaviour is only observed in crystals whose unit cells do not have a centre of symmetry, *e.g.*, the zinc blende form of ZnS.

When crystals contain permanent dipoles as a result of unit cells not having a centre of symmetry, and the dipoles are uniformly oriented in domains (like ferromagnets), they can exhibit *ferroelectric* behaviour. This means that the polarisation of the crystal increases with increasing electric field, and when that field is removed, there is a permanent polarisation and the materials can have very large dielectric constants. $BaTiO_3$, which has a distorted perovskite structure between 5 and 120°C, exhibits this behaviour because the close-packed arrangement formed by the Ba^{2+} and O^{2-} ions generates octahedral 'holes' rather too large for the Ti^{4+}. The titanium ion can rattle within the O_6 cage and so is easily polarised towards one O atom. The effect in the extended solid phase is domain polarisation. $BaTiO_3$ is used extensively in the manufacture of capacitors with very large capacitance and also has useful non-linear optical properties (frequency doubling).

Ferrites, $M^{II}Fe_2O_4$, are magnetic materials of great importance in the solid-state electronics industry. They include the inverse spinel Fe_3O_4 in which both metal ions are high-spin. Ferrites are used in cores of high-frequency transformers and in computer memory systems. More complex species such as $BaFe_{12}O_{19}$ are used to construct permanent magnets, and the yttrium iron garnet $Y_3Fe_5O_{12}$ (YIG) acts as a microwave filter in radar equipment.

A very important variant is the *inverse spinel* structure, $B[AB]O_4$, in which half the B ions are in the tetrahedral sites and the other half are in the octahedral sites together with the A ions. Such a situation occurs when the A cations have a stronger preference for octahedral coordination than do the B cations. It seems that the majority of compounds of the type $M^{IV}M^{II}_2O_4$ (*e.g.*, $Zn[ZnTi]O_4$) and many of the type $M^{II}M^{III}_2O_4$ (*e.g.*, $Fe^{III}[Co^{II}Fe^{III}]O_4$, $Fe[Fe^{II}Fe^{III}]O_4$ and $Fe^{III}[Ni^{II}Fe^{III}]O_4$) are inverse spinels.

Site selection in spinels and other systems.

A simple explanation of preference of M^{II} and M^{III} in the spinels for octahedral or tetrahedral sites can be provided by crystal field theory.

Most Cr^{III} spinels adopt a normal spinel structure (Table 2.2), the reason being due mainly to the very high octahedral site preference energy of the d^3 metal ion in the relatively weak oxide crystal field. The same is true for Mn_3O_4 which contains d^5 Mn^{2+}, which has no site preference, and d^4 Mn^{3+}, which does (octahedral). The cobalt oxide Co_3O_4 is a special case since Co^{III} is spin-paired. Since there is substantial CFSE gained from a low-spin d^6 configuration, the Co^{3+} ion prefers an octahedral site. Because the energy difference between Co^{II} in a tetrahedral and an octahedral environment is not very significant, Co_3O_4 adopts a normal spinel structure.

However, a typical example of an *inverse* spinel is $NiFe_2O_4$ in which Fe^{3+} is high spin because of the weak crystal field exerted by the oxide lattice. There is no net crystal field stabilisation for a d^5 metal ion in this symmetry, and so Fe^{III} is relatively indifferent to site location. However, d^8 Ni^{2+} has a significant octahedral preference, and the compound adopts the inverse arrangement. Magnetite, Fe_3O_4, also has an inverse spinel structure, again because Fe^{3+} has no site preference, and while d^6 Fe^{2+} has only a small octahedral preference, it is obviously sufficient to tip the balance from normal to inverse.

Table 2.2 Octahedral site preference energy (OSPE) and site occupancy in spinels

M^{III}	CFSE(tet)[a]	CFSE(oct)	OSPE[b]	Normal	Inverse
d^3	3.55(8)	12	8.45	$M^{II}Cr_2O_4$, where M^{II} = Mg, Mn, Fe, Co, Ni, Cu	
d^4	1.78(4)	6	4.22	$ZnMn_2O_4$; Mn_3O_4	$MnFe_2O_4$
d^5	0	0	0	$ZnFe_2O_4$	$M^{II}Fe_2O_4$, M^{II} = Mn, Fe, Co, Ni
d^6	2.67(6)	4	1.33	Co_3O_4	

[a] Crystal field stabilisation energy for tetrahedron calculated in units of Dq_o (Tables 1.5 and 1.6) using $Dq_t = 4/9 Dq_o$; [b] OSPE = CFSE (oct)–CFSE(tet).

While there appears to be a very good agreement between structural prediction based on CFT and the actual structures adopted by individual spinels, it should be remembered that CFSE accounts for only a small percentage of the total bonding energy of the system.

3   Metals in solution

3.1   Stability of metal complexes in solution

For a species in solution, a knowledge of its solid state structure, while being invaluable, does not provide definitive information as to what happens to that species when it is dissolved in solution. Dissociation of some or all of the ligands may occur, there may be a reaction between the ligands or the metal and the solvent, and aggregation to give di- or oligo-nuclear species could occur. For this reason, a knowledge of the stability of complexes in solution is essential.

The *thermodynamic stability* of a complex can be revealed by an equilibrium constant relating its concentration to the concentration of all other related species when the system has reached equilibrium. The *kinetic stability* of a species refers to the speed at which transformations leading to the attainment of equilibrium will occur.

> For an introduction to and basic information concerning thermodynamic stability and stability constants, see M J Winter, *d-Block chemistry*, Oxford Chemistry Primer 27, 1994, Chap. 2).

Thermodynamic stability

Providing that only mononuclear species $[M(H_2O)_{6-n}L_n]^{z+}$ are formed in a solution containing aquated metal ions $[M(H_2O)_6]^{z+}$ and unidentate ligands L, then the system at equilibrium can be described by either stepwise or overall stability constants. The former, K_n, represent six stages in the progressive substitution of H_2O by L in a six-coordinate hexa-aqua complex. To calculate the concentration of the final product, $[ML_6]^{z+}$, the overall stability constant, β, is used. The relationship between stepwise and overall formation constants is $\beta_n = K_1 K_2 K_3 ... K_n$.

> Stepwise stability constant, K_n, for reaction
> $$ML_{n-1} + L \rightleftharpoons ML_n$$
> $$K_n = \frac{[ML_n]}{[ML_{n-1}][L]}$$
> Overall stability constant, β_n, for overall reaction
> $$M + nL \rightleftharpoons ML_n$$
> $$\beta_n = \frac{[ML_n]}{[M][L]^n}$$

The chelate effect

Metal complexes of chelating ligands have significantly higher stability than their analogues containing only mono- or uni-dentate ligands. The effect can be seen clearly in the relative stabilities of $[Ni(NH_3)_6]^{2+}$ and $[Ni(en)_3]^{2+}$ (en = 1,2-diaminoethane).

$$[Ni(H_2O)_6]^{2+} + 6NH_3 \rightleftharpoons [Ni(NH_3)_6]^{2+} + 6H_2O \quad \log_{10}\beta_6 = 8.6$$

$$[Ni(H_2O)_6]^{2+} + 3en \rightleftharpoons [Ni(en)_3]^{2+} + 6H_2O \quad \log_{10}\beta_3 = 18.3$$

The origin of the chelate effect lies in enthalpic and entropic factors, the latter being dominant. The former arises mainly from two effects: electrostatic and solvation. When two monodentate ligands are brought to adjacent sites at the metal centre, they experience significant repulsion which must be overcome for substitution to take place and new bonds to form. In a chelating ligand, the coordinating centres do not have to be brought together

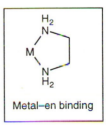

Metal–en binding

– they are already attached – so this repulsive effect is diminished. Significant solvation changes will also occur since each of the species in the equilibrium are heavily solvated. In water we would expect NH_3 to be much more highly solvated than 1,2-diaminoethane (en), and the desolvation of the former would be unfavourable enthalpically in comparison to the former.

The major effect, as stated, is the entropy change. The behaviour of NH_3 and en as ligands provides a good example since their electronic properties are virtually identical. After dissociation of one NH_3 from the complex the probability of it returning to its parent complex is virtually nil. In contrast, dissociation of one of the NH_2 groups of en would be less catastrophic, since this 'free' portion is a very short distance away from its host metal, being attached by the other end. Hence the probability of it returning to the host site is very high. Probably the most revealing picture of the entropic effect is given by the reaction

$$[Ni(NH_3)_6]^{2+} + 3en \rightleftharpoons [Ni(en)_3]^{2+} + 6NH_3$$

For this equilibrium, $\Delta G^o = -54$ kJ mol^{-1}, $\Delta H^o = -29$ kJ mol^{-1} and $\Delta S^o = +88$ J K^{-1} mol^{-1}. That the substitution will proceed is indicated by the negative value of ΔG^o, but the high positive value of ΔS^o is due principally to the conversion of four reactants ($[Ni(NH_3)_6]^{2+}$ and 3 en molecules) into seven products. There is a substantial increase in the 'disorder' in the solution, and the large value of ΔS^o is not just due to a statistical effect: the solvation shells around $[Ni(NH_3)_6]^{2+}$ and the diaminoethane preserve significant short-range ordering around these species in solution, and destruction of this contributes to the entropy increase.

Other factors affecting thermodynamic stability

The effect on stability of metal complexes as a function of the *radius* of and the *charge* on the central metal ion cannot easily be separated. Since the formation of a complex may be regarded as an interaction between a cation and either an anion or the negative end of a dipole, it is reasonable to expect that the magnitudes of the charges are important as well as the relative sizes of the interacting species. Clearly, the smaller the interacting species, or the larger the charge of, say, one cation relative to another, the greater will be the electrostatic attraction and the more stable the resulting complex (Table 3.1).

Ligands containing O or N donor atoms, *e.g.*, H_2O, OH^-, CO_3^{2-}, $O_2C_2O_2^{2-}$ (oxalate, ox^{2-}), SO_4^{2-}, NH_3 and amines, tend to form very stable complexes with Mn^{2+}, Cr^{3+} and metals in high oxidation states (IV and higher). Ligands such as PR_3, SR_2, and CO tend to form rather less stable complexes, but greatly stabilise Cu^+ and metals in low oxidation states (I, 0, –I, *etc.*). The former group of ligands contain relatively small donor atoms which are electronegative and not easy to polarise. The metals with which these ligands form the most stable complexes are also electropositive, small and very difficult to polarise. The ligands are referred to as *hard bases* and the metals as *hard acids*. The other group of ligands contain large and easily polarisable donor atoms, and are capable of entering into π bonding with metals, and the metals are also large and polarisable: these are *soft bases* and *soft acids*, respectively. A classified selection is given in Table 3.2.

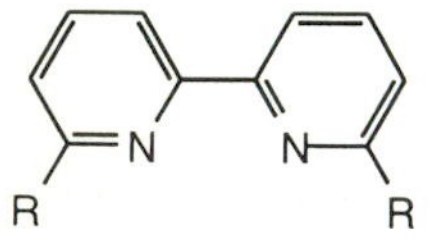

Key thermodynamic expressions relating to stability under standard conditions
$\Delta G^o = \Delta H^o - T\Delta S^o$
$\Delta G^o = -RT \ln K$
(K = stability constant)

Table 3.1 Effects of charge on stability of high-spin EDTA complexes

Metal ion	Radius	log β
V^{2+} (d^3)	0.93	12.7
V^{3+} (d^2)	0.78	25.9
Fe^{2+} (d^6)	0.92	14.3
Fe^{3+} (d^5)	0.79	25.1

Data from R D Shannon, *Acta Cryst.*, 1976, **A32**, 751.

Fig. 3.1 3,3'-R$_2$bipy: R = H, no steric hindrance; R = Me, significant steric hindrance

Steric effects can be caused by ligand substituents, for example as shown by the dipyridyl species in Fig. 3.1. The order of the first stepwise stability constants for the formation of the tetra-aqua complexes $[Fe(H_2O)_4(3,3'-R_2bipy)]^{2+}$, is K_1 (R = H) $\gg$ K_1 (R = Me). The stability of the species with R = Me is lower because the steric repulsion exerted by the methyl groups prevents close approach of the ligand to the metal centre. Steric effects in tertiary phosphine and arsine complexes are also significant. For example, $[Ni(PPh_3)_4]$ is unstable with respect to $[Ni(PPh_3)_3]$ (Section 5.3).

Normally, stepwise stability constants decrease with each successive substitution, largely for statistical reasons. However, a major change in this trend can be brought about by a change in the *spin state* of the metal ion. Dipyridyl reacts in steps with $[Fe(H_2O)_6]^{2+}$ giving *high-spin* $[Fe(H_2O)_4(bipy)]^{2+}$ $(t_{2g}^4 e_g^2)$ and then *low-spin* $[Fe(bipy)_3]^{2+}$ (t_{2g}^6), for which $K_3 \gg K_2$.

The values of stepwise stability constants for the formation of complexes of M^{2+} generally increase steadily from Mn^{2+} to Cu^{2+} and then drop at Zn^{2+}. This is the so-called 'natural' order, often referred to as the *Irving–Williams order* following the work of the discoverers of the trends. A major deviation from this order can occur in Cu^{II} chemistry, as exemplified by the reactions of $[Cu(H_2O)_6]^{2+}$ with 1,2-diaminoethane. The successive stepwise formation constants, K_1 and K_2, are normal in that they fit the natural order, and their values are greater than those of the comparable Ni^{II} complexes. However, K_3 for the reaction $[Cu(H_2O)_2(en)_2]^{2+} + en \rightleftharpoons [Cu(en)_3]^{2+}$ is the lowest of the ions Mn to Zn. This lack of stability is due to the distortion inherent in octahedral d^9 Cu^{II} – a manifestation of the *Jahn–Teller effect* (Chapter 1, p.16) – which means that whereas tetragonal distortion can readily occur in *trans*-$[Cu(H_2O)_2(en)_2]^{2+}$, a similar effect in $[Cu(en)_3]^{2+}$ would lead to unacceptable strain in at least two chelate rings. Alternatively, one could view the constraints imposed by three chelate rings as imposing near octahedral symmetry, thereby denying the complex the stabilisation inherent in Jahn–Teller distortion.

Kinetic stability: labile and inert complexes

The section above dealt with the probability of the existence of a compound or complex within a thermodynamic equilibrium. This provided no information about how quickly this equilibrium might be established, nor whether the compound will remain intact when it is formed.

Inert and labile transition metal complexes are associated with specific $3d$ orbital configurations. Inert compounds are formed by octahedral metal ions with the configurations $(t_{2g})^3(e_g)^0$, $(t_{2g})^4(e_g)^0$, $(t_{2g})^5(e_g)^0$ and $(t_{2g})^6(e_g)^0$ which give rise to very high CFSEs. Labile complexes are formed by metal ions having electrons in the e_g shell, and also with metal ions having only one or two electrons in the t_{2g} level. There are two simple reasons for this:

(i) electrons in the (antibonding) e_g orbitals of an octahedral complex repel the σ bonding electrons of the ligands;

(ii) octahedral metal ions having an empty t_{2g} orbital can use that orbital to interact with an incoming nucleophile (substituting ligand) in a transition state or intermediate.

Particular kinetic stability is associated with octahedral complexes of $Cr^{III}\{(t_{2g})^3(e_g)^0\}$ and $Co^{III}\{(t_{2g})^6(e_g)^0\}$, whereas high-spin Fe^{III} or Mn^{II} $(t_{2g}^3 e_g^2)$ are labile. Tetrahedral species with $3d^6$–$3d^8$ are usually labile, except under very special circumstances.

Table 3.2 Some hard and soft acids and bases

Hard acids		Hard bases	
Ti^{4+}	VO^{2+}	NH_3 en H_2O	
Cr^{6+}	Cr^{3+}	OH^-	CO_3^{2-}
Mn^{7+}	Mn^{2+}	SO_4^{2-}	F^-
Fe^{3+}	Co^{3+}	PO_4^{3-}	Cl^-

Soft acids	Soft bases	
Cu^+, metals in	H^- alkyl aryl	
oxidation state	CN^- CO	
0, –1, etc.	alkene SR^-	
	SR_2 PR_3 I^-	

Borderline cases		
Br^- N_3^- N_2	Fe^{2+} Co^{2+}	
NO_2^-	Ni^{2+} Cu^{2+}	

CFSE ≡ crystal field stabilisation energy.

If the process of substitution of one ligand by another at a metal centre is complete in less than one minute (at RT with *ca.* 0.1 M solutions) then the complex undergoing reaction is regarded as *labile*. If the substitution takes longer than this, then it is regarded as *inert*. Inert complexes are not necessarily thermodynamically stable with respect to a particular reaction, and thermodynamically stable complexes can undergo rapid reactions.

3.2 Electrode potentials and redox reactions

Transition metal ions are unique in their ability to exist in several different oxidation states, to interconvert between those oxidation states in single or multiple electron transfer steps, and to participate in highly selective redox processes. The relative thermodynamic stabilities of oxidation states of given metals in different environments (solvents, effect of ligand type), the power of metals and their compounds as oxidising or reducing agents, and the capability of complexes to engage in electron transfer reactions (redox processes) are described by electrochemical terms and measured in terms of electrode potentials. While full discussion of electrochemistry is beyond the scope of this Primer, it is necessary to mention electrode potentials and half-cell reactions.

Electrochemical properties in aqueous media

The study of the electrochemical properties of inorganic compounds developed originally in water, and much thermodynamic data connected with redox activity was referred to reactions in aqueous media. The reference system for electrode potentials in water is the *standard hydrogen electrode* (SHE). The half-cell reaction is shown as

$$H_2 \rightleftharpoons 2H^+(aq) + 2e^-$$

> E^o is defined as the potential of a couple (M^{n+}/M) under *standard* conditions (25°C, 1 stm. pressur, unit concentration): E is the *measured* potential in any cell).

the potential, E^o, being defined as 0.00 V under standard (thermodynamic) conditions. Electrochemical information is often obtained in other solvents (*e.g.*, CH_2Cl_2), and other reference electrodes are used such as the saturated calomel electrode (SCE, Hg^+/Hg) and the AgCl/Ag electrode. However, the potentials quoted against these references and in solvents other than water are not 'standard' potentials: they can be described as *formation* potentials.

In a cell consisting of a SHE and a metal electrode, M, dipped in an aqueous solution containing M^{n+}, the standard potential of the electrode is the same as the measured potential of the cell since the SHE potential is defined as 0.00 V under standard conditions. If the metal electrode is positively charged with respect to the hydrogen electrode, then the electrode potential is positive, as in the half-reaction $Cu^{2+} + 2e^- \rightarrow Cu$ where $E = +0.34$ V. If the metal electrode becomes negatively charged (it loses electrons more readily than hydrogen), the electrode potential is negative, as in the half-reaction $Mn^{2+} + 2e^- \rightarrow Mn$ where $E = -1.18$ V. These electrode potentials reflect the electrostatics of what is happening at the metal electrode, and are invariant quantities irrespective of the type of electrochemical cell in which they are measured.

> Electrode potentials are measured in cells consisting of a test or working electrode such as Hg, Pt, or Au, and a reference electrode such as the SHE or the saturated calomel electrode (SCE, Hg^+/Hg) in aqueous media or AgCl/Ag in non-aqueous media. For more details see P. W. Atkins, *Physical chemistry*, 3rd Ed., OUP, 1986, Chap.11.

However, for the purposes of discovering whether a particular redox reaction will occur, it is more useful to develop a thermodynamic convention involving potential E^o, defined in the relationship

$$\Delta G^o = -nF\Delta E^o$$

where ΔG^o is the change in the standard *Gibbs free energy*, n is the number of moles of electrons involved in the reaction and F is Faraday's constant. In this convention it is necessary to indicate the direction of the reaction.

We know that the reaction $Ni + 2H^+ \rightarrow Ni^{2+}(aq) + H_2$ occurs readily. This means that $\Delta G^o < 0$ and so, since the SHE is defined as 0.00 V ($\Delta G^o = 0$), for the half-reaction $Ni \rightarrow Ni^{2+}(aq) + 2e^-$, $\Delta E^o > 0$. For the non-spontaneous reaction $Ni^{2+}(aq) + H_2 \rightarrow Ni + 2H^+$, $\Delta G^o > 0$ and for the half-reaction $Ni^{2+}(aq) + 2e^- \rightarrow Ni$, $\Delta E^o < 0$.

Here potential E is used exclusively to report redox properties, and half-reactions are reported as *reduction processes*. Some standard reduction potentials for first-row transition metal ions in aqueous acid are listed in Table 3.3.

Before it is possible to discuss the relative stabilities of oxidation states of particular metals, several other points must be made:

(a) For the general reaction $xOx_A + yRed_B \rightarrow aRed_A + bOx_B$, the Nernst equation applies.

$$E = E^o - \frac{RT}{nF} \log Q, \text{ where } Q = \frac{[Red_A]^a [Ox_B]^b}{[Ox_A]^x [Red_B]^y}$$

> The sign of E^o for a half-reaction or an overall redox reaction depends on the direction in which the reaction is written, and this is entirely consistent for other thermodynamic quantities such as enthalpy and entropy. Convention requires that half-reactions are reported as reduction processes.

(b) Reactions resulting in a decrease in Gibbs free energy are spontaneous, and so redox reactions with $E^o > 0$ are also spontaneous.

(c) In aqueous solution there are two reactions of particular importance which limit the thermodynamic stability of species in aqueous solution: (i) the reduction of hydrogen in water or in H_3O^+, and (ii) the oxidation of oxygen in water or hydroxide ions.

(i) $H_3O^+ + e^- \rightarrow H_2O + 1/2H_2$ (1M acid) $E^o = 0.00V$
    $H_2O + e^- \rightarrow OH^- + 1/2H_2$ (neutral soln) $E^o = -0.41V$
    $H_2O + e^- \rightarrow OH^- + 1/2H_2$ (1M base) $E^o = -1.23V$

(ii) $1/2O_2 + 2H^+ + 2e^- \rightarrow H_2O$ (1M acid) $E^o = -1.23V$
    $1/2O_2 + 2H^+ + 2e^- \rightarrow H_2O$ (neutral soln) $E^o = -0.19V$
    $1/2O_2 + H_2O + 2e^- \rightarrow 2OH^-$ (1M base) $E^o = -0.40V$

Taking these half-reactions into account, it is relatively easy to discover from Table 3.3 which oxidation states of individual metals are thermodynamically stable in water in acid conditions. Potential data can also be measured in basic solution, but are not tabulated here.

The stability of the highest oxidation states of the elements decreases markedly on passing from Ti to Mn. The reduction potentials in aqueous acid confirm this. Iron in oxidation state VI is particularly unstable, and is not encountered in aqueous solution, although ferrates, $[FeO_4]^{2-}$, may be stabilised in the solid state. Cobalt(III) is unstable in water, as is Ni(IV), although, like Fe(VI), these relatively high oxidation state species are found in binary and ternary oxides and related solids. Conversely, low oxidation states are almost never encountered unless π-acceptor ligands are bound to the ion. The exception is Cu(I), whose stability is very much influenced by the co-ligands (see below).

> Reduction potentials may vary significantly in non-aqueous solvents, leading to a marked change in stability of a particular metal oxidation state.

Broadly speaking, the reduction potentials for the first-row transition metals in aqueous acid confirm the decreasing stability of the highest oxidation states of each metal, reaching a maximum at Fe^{VI}. The same is true in basic solution, although the highest oxidation states are slightly less oxidising in this medium (see below).

Table 3.3 Standard reduction potentials in aqueous acid solution

Electrode	Oxidation state changes	E^O (V)
$Ti^{2+} + 2e^- \rightarrow Ti$	Ti(II) $\rightarrow$ Ti(0)	−1.63
$TiO^{2+} + 2H^+ + 4e^- \rightarrow Ti + H_2O$	Ti(IV) $\rightarrow$ Ti(0)	−0.88
$Ti^{3+} + e^- \rightarrow Ti^{2+}$	Ti(III) $\rightarrow$ Ti(II)	−0.37
$TiO^{2+} + 2H^+ + e^- \rightarrow Ti^{3+} + H_2O$	Ti(IV) $\rightarrow$ Ti(III)	+0.10
$V^{2+} + 2e^- \rightarrow V$	V(II) $\rightarrow$ V(0)	−1.19
$V^{3+} + e^- \rightarrow V^{2+}$	V(III) $\rightarrow$ V(II)	−0.26
$V(OH)_4^+ + 2H^+ + e^- \rightarrow VO^{2+} + 3H_2O$	V(V) $\rightarrow$ V(IV)	+1.00
$Cr^{3+} + 3e^- \rightarrow Cr$	Cr(III) $\rightarrow$ Cr(0)	−0.74
$Cr^{3+} + e^- \rightarrow Cr^{2+}$	Cr(III) $\rightarrow$ Cr(II)	−0.41
$Cr_2O_7^{2-} + 14H^+ + 6e^- \rightarrow 2Cr^{3+} + 7H_2O$	Cr(VI) $\rightarrow$ Cr(III)	+1.33
$Mn^{2+} + 2e^- \rightarrow Mn$	Mn(II) $\rightarrow$ Mn(0)	−1.18
$MnO_4^- + e^- \rightarrow MnO_4^{2-}$	Mn(VII) $\rightarrow$ Mn(VI)	+0.56
$MnO_2 + 4H^+ + 2e^- \rightarrow Mn^{2+} + 2H_2O$	Mn(IV) $\rightarrow$ Mn(II)	+1.23
$Mn^{3+} + e^- \rightarrow Mn^{2+}$	Mn(III) $\rightarrow$ Mn(II)	+1.51
$MnO_4^- + 8H^+ + 5e^- \rightarrow Mn^{2+} + 4H_2O$	Mn(VII) $\rightarrow$ Mn(II)	+1.51
$MnO_4^- + 4H^+ + 3e^- \rightarrow MnO_2 + 2H_2O$	Mn(VII) $\rightarrow$ Mn(IV)	+1.70
$Fe^{2+} + 2e^- \rightarrow Fe$	Fe(II) $\rightarrow$ Fe(0)	−0.44
$Fe^{3+} + e^- \rightarrow Fe^{2+}$	Fe(III) $\rightarrow$ Fe(II)	+0.77
$FeO_4^{2-} + 8H^+ + 3e^- \rightarrow Fe^{3+} + 4H_2O$	Fe(VI) $\rightarrow$ Fe(III)	+2.20
$Co^{2+} + 2e^- \rightarrow Co$	Co(II) $\rightarrow$ Co(0)	−0.28
$Co^{3+} + e^- \rightarrow Co^{2+}$	Co(III) $\rightarrow$ Co(II)	+1.81
$Ni^{2+} + 2e^- \rightarrow Ni$	Ni(II) $\rightarrow$ Ni(0)	−0.25
$NiO_2 + 4H^+ + 2e^- \rightarrow Ni^{2+} + 2H_2O$	Ni(IV) $\rightarrow$ Ni(II)	+1.68
$Cu^{2+} + e^- \rightarrow Cu^+$	Cu(II) $\rightarrow$ Cu(I)	+0.15
$Cu^{2+} + 2e^- \rightarrow Cu$	Cu(II) $\rightarrow$ Cu(0)	+0.34
$Cu^+ + e^- \rightarrow Cu$	Cu(I) $\rightarrow$ Cu(0)	+0.52

Standard reduction potentials in basic solution:
(i) $MnO_4^- + 2H_2O + 3e^- \rightarrow MnO_2 + 4OH^-$ $\qquad E^O = +0.59$ V
(ii) $Co(OH)_3 + e^- \rightarrow Co(OH)_2 + OH^-$ $\qquad E^O = +0.17$ V

However, the potential data reveal the opposite trends in the stabilities of the lower oxidation states, particularly III and II, with the exceptional stability of Cu^I being accounted for by its electronic configuration (d^{10}). The observed reduction potentials for the reaction $[M(H_2O)_6]^{3+} \rightarrow [M(H_2O)_6]^{2+}$ broadly follow the trends in third ionisation potential ($M^{2+} \rightarrow M^{3+}$), the discrepancies being partly accounted for by size of the respective ions and by covalency in the M–O bonds.

$[Ti(H_2O)_6]^{3+}$ is relatively easily obtained by chemical or electrolytic reduction of aqueous Ti(IV) species, but reduces oxygen very efficiently and so must be handled under nitrogen or argon. As a reducing agent it is not as apparently powerful as $[Cr(H_2O)_6]^{2+}$ but it reacts at a comparable rate in acidic solutions, and is generally a kinetically fast reductant. Vanadium(V) is moderately strongly oxidising, but can be reduced using mild agents to give the vanadyl(IV) ion, $[VO(H_2O)_5]^{2+}$. The vanadyl ion is also readily generated by reaction of $[V(H_2O)_6]^{3+}$ in solution with air. Solutions containing $[V(H_2O)_6]^{2+}$ are extremely air sensitive, but can reduce water to hydrogen, even though the potentials seem to suggest that this would be unlikely.

Chromium(VI) is powerfully oxidising, the dichromate ion, $Cr_2O_7^{2-}$, being particularly effective in acid solution, the chromate ion, CrO_4^{2-}, being less so in basic media. Chromium(III) is the product from both ions. Reduction of $[Cr(H_2O)_6]^{3+}$ affords air-sensitive $[Cr(H_2O)_6]^{2+}$ which can be used as a mild reducing agent. The most stable oxidation state of manganese in aqueous media is Mn(II), but $[Mn(H_2O)_6]^{3+}$ can be generated using strong oxidising agents, or by controlled reduction of MnO_4^-. The highest oxidation states, Mn(VI) and Mn(VII), are very strongly oxidising, and consequently very unstable. The manganate ion, MnO_4^{2-}, is only stable in basic solution, readily disproportionating in neutral, acidic, or even weakly basic media to give MnO_2 and MnO_4^- ($3Mn^{VI} \rightarrow Mn^{IV} + 2Mn^{VII}$).

$[Fe(H_2O)_6]^{3+}$ is easily obtained by oxidation of $[Fe(H_2O)_6]^{2+}$, both ions being relatively stable in aqueous solution when the pH is carefully controlled (Section 4.1). In the absence of coordinating groups, oxidation of $[Co(H_2O)_6]^{2+}$ to $[Co(H_2O)_6]^{3+}$ is very unfavourable but aquated Co(III) is much more stable in alkaline solution. Nickel(II) is the most stable oxidation state in aqueous media, and higher and lower oxidation states are only stabilised by highly electronegative or π-acceptor ligands, respectively. Cu(I) in aqueous media is stabilised with respect to Cu(II) by CN^- in excess (as $[Cu(CN)_2]^-$) or I^- (ligands capable of acting as good σ donors and as a π acceptor or donor, respectively).

Redox properties of metal complexes

The effect of ligands in complexes on the electrode potential for a couple $[ML_n]^{z+1}/[ML_n]^z$ is significant, as shown by the data obtained from cobalt complexes (Table 3.4).

Table 3.4 Reduction potentials for the Co^{III}/Co^{II} redox couple

Reduction reaction	E^O (V)
$[Co(H_2O)_6]^{3+} + e^- \rightarrow [Co(H_2O)_6]^{2+}$	+1.84
$[Co(EDTA)]^- + e^- \rightarrow [Co(EDTA)]^{2-}$	+0.60
$[Co(ox)_3]^{3-} + e^- \rightarrow [Co(ox)_3]^{4-}$	+0.57
$[Co(phen)_3]^{3+} + e^- \rightarrow [Co(phen)_3]^{2+}$	+0.42
$[Co(NH_3)_6]^{3+} + e^- \rightarrow [Co(NH_3)_6]^{2+}$	+0.10
$[Co(en)_3]^{3+} + e^- \rightarrow [Co(en)_3]^{2+}$	−0.26
$[Co(CN)_6]^{3-} + e^- \rightarrow [Co(CN)_5(H_2O)]^{3-} + CN^-$	−0.83

As mentioned above, the data relating to the reduction $[Co(H_2O)_6]^{3+} \rightarrow [Co(H_2O)_6]^{2+}$ in aqueous acidic solution reveal that Co(III) is very unstable, oxidising water readily. However, by replacing water with other ligands, the relative stability of Co^{III} increases dramatically. The reason for this is partly related to the crystal field strength of the ligand and concomitant change in spin state, from the very stable $t_{2g}{}^6 e_g{}^0$ arrangement in low-spin Co^{III} to $t_{2g}{}^5 e_g{}^2$ in Co^{II}, passing through the unstable intermediate step of $t_{2g}{}^6 e_g{}^1$. It is also probable that entropy changes associated with the presence of chelating ligands in some of the complexes account for small deviations from the order of the spectrochemical series ($ox^{2-} < H_2O < NH_3 < en < phen < CN^-$).

Ligand effects are also observed in the redox potentials of iron complexes (Table 3.5). In all these complexes, the lower oxidation state is more stable, but the order does not apparently parallel the crystal field strength of the ligand. The complexes containing phen and CN^- are low spin in both Fe^{III} and Fe^{II}, and the configuration in the latter, $t_{2g}{}^6 e_g{}^0$, is particularly stable. However, both aqua complexes are high spin (Fe^{III} $t_{2g}{}^3 e_g{}^2$, and Fe^{II} $t_{2g}{}^4 e_g{}^2$), as is $[Fe(oxin)_3]^z$ (Fig. 3.2). It appears that anions stabilise Fe^{III} with respect to Fe^{II}, which is probably an entropic rather than an enthalpic effect. The neutralisation of charge on a cationic complex by combination with an anion increases the entropy of the whole system. This occurs because the highly organised solvation shell at the cation is disrupted on neutralisation with release of the previously organised solvent into the bulk solution (*i.e.*, disorder increases).

The reduction potential for the reaction $[Fe(CN)_6]^{3-} \rightarrow [Fe(CN)_6]^{4-}$ is the most relatively positive of the anionic species, indicating the increasing stability of Fe^{II} relative to Fe^{III} as σ-donor ligands are replaced by good π acceptors. A similar explanation can be offered for the very positive reduction potential for the reduction of $[Fe(phen)_3]^{3+}$ to $[Fe(phen)_3]^{2+}$ compared to their hexa-aqua analogues; namely that while both phen and H_2O are good σ donors, only the former is a very good π acceptor. In addition, low spin Fe^{II} is a particularly good π donor, better than Fe^{III} with its higher charge and very high ionisation potential to Fe^{4+}.

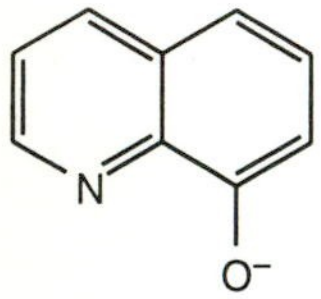

Relative oxidation state stability in a given metal complex, *e.g.*, $[Fe(5,5'-R_2bipy)_3]^{n+}$ (Fig. 3.2), can also be influenced by the relative basicity and/or π acceptor ability of a ligand such as 2,2'-dipyridyl. In general, electron-releasing substituents, such as Me, OMe, and NMe_2, will destabilise Fe^{II} (by increasing the ligands basicity), whereas substituents having an electron withdrawing effect, *e.g.*, CN, CO_2R, or NO_2, will increase the relative stability of Fe^{II}, partly as a result of reduced basicity but also because of increased π-acceptor capability.

It is not possible to say that a particular ligand will invariably stabilise a given oxidation state relative to another which is higher or lower. However, it is broadly true to say (i) that the low oxidation states will be stabilised by electron-acceptor ligands (Chapter 5), and (ii) that high oxidation states will be stabilised by electron-donor ligands (Chapter 4). However, specific stable electron configurations, *e.g.* $(t_{2g})^6$, may provide exceptions.

Table 3.5 Reduction potentials of iron complexes

$[Fe(phen)_3]^{3+} \rightarrow [Fe(phen)_3]^{2+}$
$E^o = +1.12$ V
$[Fe(H_2O)_6]^{3+} \rightarrow Fe(H_2O)_6]^{2+}$
$E^o = +0.77$ V
$[Fe(CN)_6]^{3-} \rightarrow [Fe(CN)_6]^{4-}$
$E^o = +0.36$ V
$[Fe(EDTA)]^- \rightarrow [Fe(EDTA)]^{2-}$
$E^o = -0.12$ V
$[Fe(oxin)_3] \rightarrow [Fe(oxin)_3]^-$
$E^o = -0.20$ V

8-Hydroxy-quinolate (oxin)

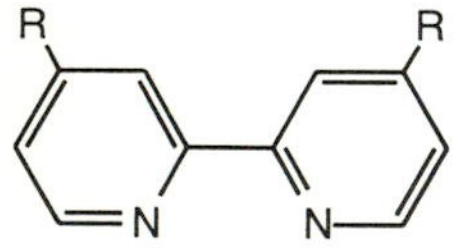

5,5'-R_2bipy

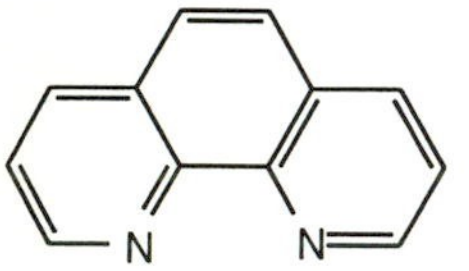

phen

Fig. 3.2 Ligands mentioned in Tables 3.4 and 3.5

3.3 Mechanism of substitution

The ways in which ligands are replaced within a metal coordination sphere are summarised here, dealing first with a classification of mechanisms, then with substitution in octahedral, planar, and tetrahedral complexes.

Classification of substitution mechanisms

Generally mechanisms are considered from two kinetic aspects: the *stoichiometric mechanism*, which is the sequence of simple steps by which the reaction takes place, and the *intimate mechanism*, which explores the details of the activation process and the energetics of formation of the activated complex in the rate-determining step.

Stoichiometric mechanism

There are three types of stoichiometric mechanisms: dissociative, D, associative, A, and interchange, I. The first two are two-step processes requiring the formation of an intermediate which is then converted into the product. The dissociative process involves prior departure of a leaving group leading to an intermediate of lower coordination number. Many octahedral complexes (coordination compounds like $[Co(NH_3)_5Cl]^{2+}$ and metal carbonyls, $[Cr(CO)_6]$) undergo substitution by dissociative mechanisms. The associative process involves formation of an intermediate with a higher coordination number than the original reactant, and planar complexes (particularly those of Ni^{II}) are substituted by this route. The third is a one-step concerted process, and it is difficult to detect an intermediate. Substitution reactions of $[Ni(H_2O)_6]^{2+}$ occur *via* an interchange process.

Intimate mechanism

The precise details of the formation of the activated complex define the intimate mechanism, and this is especially important in determining the nature of the interchange mechanism. A dissociatively activated reaction is one whose rate is insensitive to changes in the nature of the entering group. An associatively activated reaction is one in which the rate is sensitive to the nature of the entering group. In classifying interchange mechanisms, if the rate constant for the formation of the activated complex is significantly dependent on the entering group, it is described as I_a. If the reaction rate constant is largely independent of the entering group then it is I_d.

Substitution in octahedral complexes

Many but not all octahedral complexes, be they of the classical coordination type (involving O-, N-, S- or P-containing ligands), or metal carbonyls and other low-oxidation state species, undergo substitution *via* a dissociatively activated process.

In any substitution process, the solvent plays an extremely important role. At the simplest level, the solvent can be itself a nucleophile and as such will compete with an entering group. This is shown in Fig. 3.3. The displacement of the leaving group X by solvent is the dominant substitution characteristic of most coordination compounds in aqueous solution, but the solvent may be of less direct importance in the substitution processes of metal carbonyls and organometallic species which can occur in non-polar media.

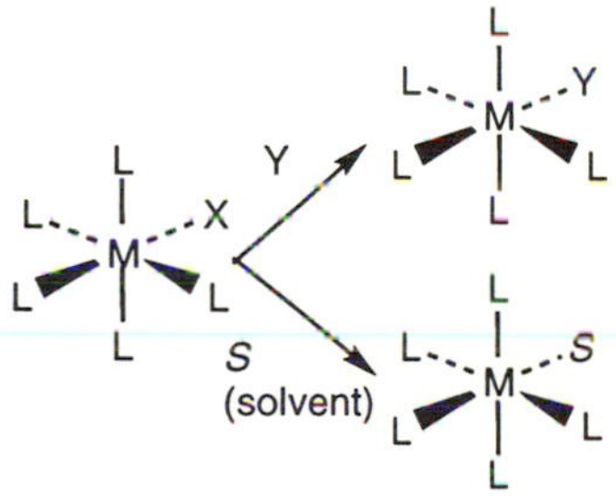

Dissociative pathways should be independent of the nature and concentration of the entering group. However, since the activation process

For an introduction to and more detailed discussion of reaction mechanism see R. A. Henderson, *The mechanisms of reactions at transition metal sites*, Oxford Chemistry Primer, 10, 1995.

There are two extremes for substitution pathways in octahedral complexes: *dissociative* and *associative*. Crystal field calculations suggest that there will be a large loss of CFSE energy for d^3, low-spin d^4, low-spin d^5 and low-spin d^6 complexes in converting from 6- to either 5- or 7-coordinate. This means that there is a significant crystal field contribution to the activation energy for the substitution process, and so metal ions with these configurations are inert. All other configurations lose little or no CFSE and so are relatively labile.

Fig. 3.3 Solvation of an octahedral complex (S = solvent, Y = nucleophile)

requires breaking the M–X bond, it can be expected that the leaving group makes a very strong contribution to the reaction rate.

A typical reaction might be the hydration of $[Co(NH_3)_5X]^{2+}$:

$$[Co(NH_3)_5X]^{2+} + H_2O \rightarrow [Co(NH_3)_5(H_2O)]^{3+} + X^-$$

In simple terms, the metal site in this reaction system may be described as 'hard', and dissociation of 'hard' ligands will be less easy than that of 'soft' ligands: the first order rate constant falls in the order $I > Br > Cl > F$. If the reaction centre is 'softened', as occurs when the metal is surrounded by π-acceptor ligands as in $[Co(CN)_5X]^{3-}$ (X = halide), then this order is reversed: $F > Cl > Br > I$.

Spectator ligand effects

The ligands around the metal centre tune the relative 'electron richness' of the metal site which, in turn, influences the dissociation of the leaving group. The general effects of the co-ligands in octahedral complexes are very much smaller than in planar complexes, and there is no dominant *trans* effect. However, substitution studies of the geometric isomers *cis-* and *trans-*$[Co(en)_2ZCl]^{n+}$ (Z = spectator ligand), show that:

(a) σ-donor or π-acceptor Z ligands labilise the Co–X bond if they are *trans* to the leaving group, by weakening the Co–Cl bond;

(b) π-electron donors, when *cis* to the leaving group, can donate electron density to a metal orbital which can also interact with the leaving group, thereby repelling the leaving group (Fig. 3.4).

An example of (a) may be the faster hydrolysis of $[Ni(NH_3)_5X]^+$ *vs.* $[Ni(H_2O)_5X]^+$ giving $[NiL_5(H_2O)]^{2+}$ (L = NH_3 or H_2O; X = halide ion). Ammonia is a stronger σ donor than water leading to an increase of electron density on the metal and greater repulsion of X from Ni. In the activated complex, good σ donors stabilise lower coordination numbers.

Effects of overall charge

It is very difficult to make comparisons of complexes where the only variable in the system is overall ionic charge. However, we might expect that since nucleophiles are either anions or the negative ends of dipoles, an increase in positive charge on the substituting complex, or a reduction in negative charge, would favour an associatively activated process. From the limited data available from cobalt(III)/NH_3 complexes, it appears that this is so, an increase of charge on the complex from +2 to +3 decreasing the rate of hydrolysis of a Co–Cl bond by a factor of between 100 and 1000 at 25°C.

Stereochemistry of substitution

Generally, dissociatively activated substitution reactions of octahedral complexes occur with retention of configuration (Scheme 3.1(a)), and this is consistent with a square pyramidal five-coordinate intermediate (Fig. 3.5).

However, the energy difference between a square pyramidal intermediate and the alternative trigonal bipyramid is very small. In $[Co(NH_3)_4(Z)Cl]^{n+}$ or $[Co(en)_2(Z)Cl]^{n+}$, where Z is a π donor, then a mixture of *cis* and *trans* isomers can be formed if Z can be incorporated into the trigonal plane of the

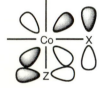

Fig. 3.4 *Cis* effect of π-donor Z on the repulsion of leaving group X; strong interaction between Z and Co will tend to weaken the Co–X bond

Fig. 3.5 Stabilisation of a trigonal bipyramidal intermediate by π bonding between Z and the metal

trigonal bipyramidal intermediate. Attack by the entering group on such an intermediate can occur at any one of three positions (Scheme 3.1(b)).

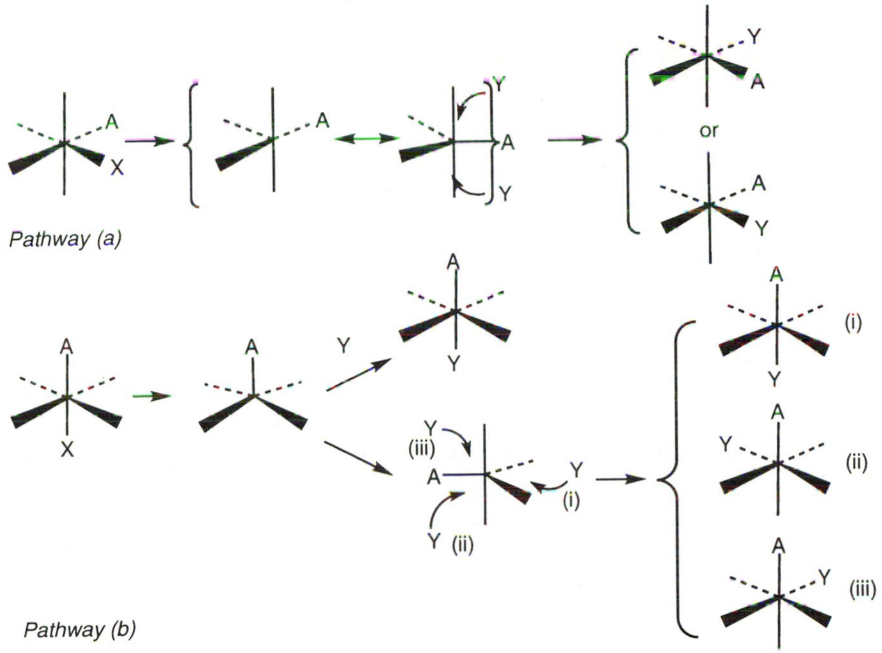

Pathway (a)

Pathway (b)

Scheme 3.1 Stereochemistry of substitution in octahedral complexes (Y = nucleophile)

Base-catalysed hydrolysis

The hydrolysis of ammine complexes, particularly of Co^{III} and Cr^{III}, is greatly accelerated by base catalysis. While the rate law for a process such as hydrolysis of $[Co(NH_3)_5Cl]^{2+}$ to $[Co(NH_3)_5(OH)]^{2+}$, is dependent on the concentration of $[Co(NH_3)_5Cl]^{2+}$ and OH^-, the mechanism is not a simple bimolecular process involving direct substitution of Cl^- by OH^-.

Hydroxide deprotonates an ammonia giving coordinated NH_2^-, this being facilitated by the binding of the NH_3 to Co^{3+} which increases the acidity of the H atoms. The NH_2^- is a powerful π donor which can stabilise the five-coordinate intermediate (Fig. 3.6). The role of OH^- is to act as a deprotonating agent, not a nucleophile, and addition of H_2O to the intermediate gives the product $[Co(NH_3)_5(OH]^{2+}$.

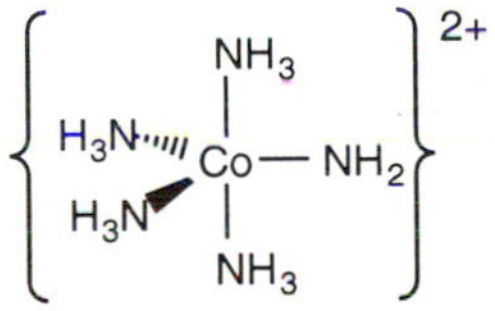

Fig. 3.6 Five-coordinate intermediate in base-catalysed hydrolysis of cobalt ammine complexes

Acid-catalysed hydrolysis

Protonation of a ligand can enable its dissociation, particularly if the ligand is a conjugate base of a relatively weak acid. Hydrolysis of the first halide in *cis*-$[Co(en)_2X_2]^+$ in aqueous acid is normally independent of the acid concentration when X = Cl, Br or I. This is partly due to the corresponding acids of the leaving group being strong. However, HF is a weak acid, and the hydrolysis of the difluoride is strongly acid catalysed, as shown in Scheme 3.2. Protonation of the fluoride changes the leaving group from an anion to a neutral species which, in the transition state, is more easily released from the cationic metal centre than an anionic group.

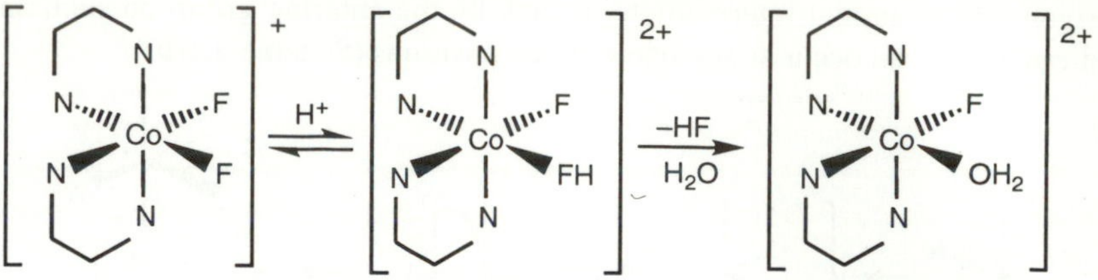

Scheme 3.2 Acid-catalysed hydrolysis of *cis*-[Co(en)$_2$F$_2$]$^+$

Isomerisation

In Scheme 3.1, we showed how substitution of *cis* and *trans* isomers *via* a dissociative route leads to isomerisation. In octahedral complexes containing only chelating ligands, isomerisation can occur also *via* a dissociative pathway in which only one donor atom detaches from the complex, thereby forming a five-coordinate intermediate. Since the energy difference between a square pyramidal and trigonal bipyramidal intermediate is very small, and once the latter has been formed, it is extremely easy to interchange axial and equatorial groups *via* pseudo-rotation (Scheme 3.3). The detached group has the opportunity to return to the metal creating either a species identical to the starting compound, or its isomer (Scheme 3.4).

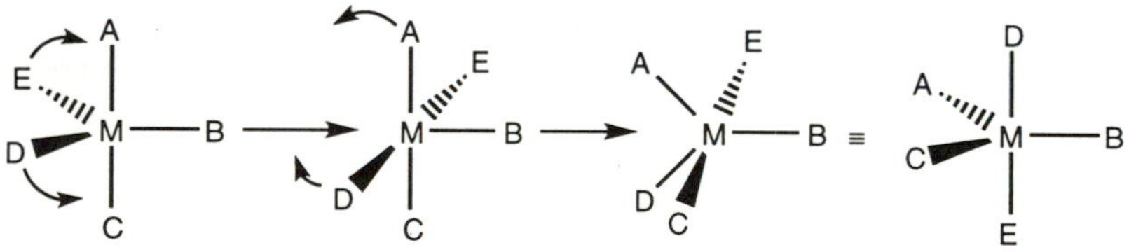

Scheme 3.3 Pseudo-rotation or Berry rotation in a trigonal bipyramidal intermediate

However, tris(chelate) complexes may also isomerise by intramolecular processes. There are two possible pathways: the Bailar or trigonal twist and the Ray–Dutt or rhombic twist (Scheme 3.5).

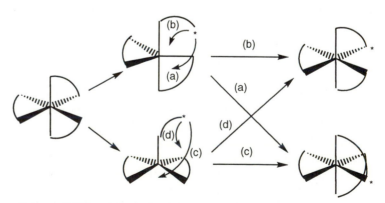

Scheme 3.4 Dissociative isomerisation pathways for tris(chelate) complexes

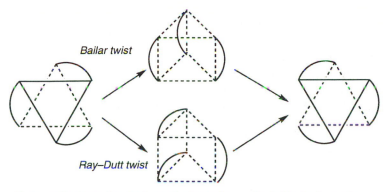

Bailar twist

Ray–Dutt twist

Scheme 3.5 Intramolecular isomerisation pathways for tris(chelate) complexes

Note that the two octahedra shown here are non-superimposable mirror images: the metal centre is chiral and the enantiomers can interconvert by a twist mechanism.

Tris(chelate) complexes exist in enantiomeric configurations about the metal atom, and when the metal centre is inert, it is possible to isolate the individual enantiomers. The interconversion of these enantiomers, leading to racemisation if the starting material is a pure enantiomer, can provide important spectroscopic information about the mechanism of isomerisation. It seems that racemisation of pure enantiomers of $[Cr(ox)_3]^{3-}$ and of the β-diketonate $[Co(MeCOCHCOPr^i)_3]$ occurs by dissociative pathways, whereas $[Co(S_2CNR_2)_3]$ racemises by a twist mechanism.

Twist mechanisms are not confined to tris(chelate) complexes, since *cis*-$[Cr(CO)_4(PR_3)_2]$ isomerises to the *trans* isomer by such a process.

$$S_2CNR_2 = \begin{array}{c} S \\ \\ S \end{array} \hspace{-0.5em} > \hspace{-0.3em} C-NR_2$$

Water exchange

The simplest substitution reaction is exchange of coordinated water around a metal ion in aqueous solution, no other ligands being present. The exchange rates for first-row transition metal ions range from about 10^9 to about 10^{-3} sec^{-1}.

High-spin Cr^{2+} (d^4) and Cu^{2+} (d^9) undergo extremely fast exchange ($k \geq 10^8$ sec^{-1}), being essentially diffusion controlled processes. The high-spin ions Mn^{2+} (d^5), Fe^{2+} (d^6), Co^{2+} (d^7) and Ni^{2+} (d^8) have k in the range $10^4 - 10^7$ sec^{-1} while V^{2+} (d^3) exchanges more slowly ($k \approx 10^2$ sec^{-1}). The ions Cr^{3+} (d^3) and Co^{3+} (d^6) are kinetically inert, having exchange rates between 10^{-3} and 10^{-6} sec^{-1}.

We should recall that *inert* complexes are those with very high CFSEs, namely d^3, low-spin d^4, low-spin d^5 and low-spin d^6. Only the first of these configurations is normally encountered in aqua complexes. Those ions having d^1 or d^2 are kinetically labile, as are Cu^{II} complexes, partly because of Jahn–Teller distortion. Charge on the ion can also be important: V^{2+} (d^3) is somewhat labile while Cr^{3+} (d^3) is inert. Octahedral complexes of V^{3+} (d^2) and Ni^{2+} (d^8) also have quite significant CFSEs, and so react more slowly than high-spin d^5-d^7 metal ions.

Substitution in planar complexes

The main examples of planar first-row transition metal complexes are encountered in d^8 Ni^{II} chemistry. In general, these complexes undergo ligand substitution by associative pathways but there is relatively little mechanistic information, studies having been made mainly of Pd^{II} and Pt^{II} analogues. It is significant, however, that five-coordinate $[Ni(CN)_5]^{3-}$ is stabilised in crystals, and this is powerful support for the associatively activated exchange of cyanide ion with planar $[Ni(CN)_4]^{2-}$.

In associatively activated reactions, the nature and concentration of the entering ligand is extremely important and, of course, the solvent can compete with the incoming nucleophile. A correlation between the nucleophilicity of the entering ligand and reaction rate can be made, but this nucleophilicity is really a function of the hardness or softness of both the metal centre and the entering group. The reaction proceeds *via* a five-coordinate intermediate in which stereochemistry is usually, but not invariably, preserved. The solvent can compete with the entering group Y, giving rise to a solvento-complex. Displacement of solvent by Y will then occur by precisely the same mechanism as that shown in Scheme 3.6.

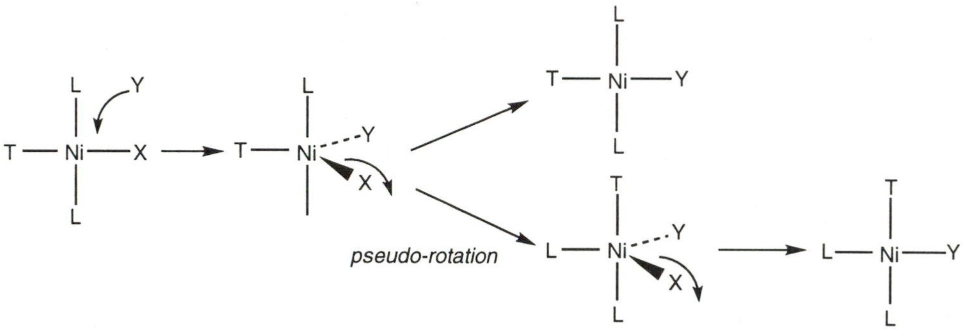

Scheme 3.6 Substitution in planar complexes showing isomerisation (T = *trans* ligand; Y = nucleophile)

A most important aspect of associatively activated substitution reactions is the role of the ligand *trans* to the leaving group, *viz.*, T. This ligand can enormously enhance the rate of substitution of the complex, and this is known as the *trans* effect. The rate of substitution increases in the order T = NH_3, py, OH^-, H_2O < Cl^-, Br^- < NO_2^-, I^-, SCN^-, Ph^- < Me^- < H^-, PR_3 < CO, CN^-, C_2H_4. The strength of the effect depends on both σ and π bonding between T and the metal: the greater the overlap between the ligand and metal orbitals in the M–T bond, the stronger the effect.

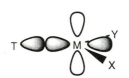

Fig. 3.7 Effect of a strong σ-bonding *trans* ligand on M–T bonding

Ligands such as H^- and Me^- are extremely electronegative: they polarise the electron density in the T–M bond thereby weakening (labilising) the M–X bond, and assisting addition of Y. This situation is illustrated in Fig. 3.7 and is mainly a ground state effect.

The ligands CO, alkenes, CN^- and PR_3 are π acceptors capable of removing surplus electron density at the metal, thus facilitating addition of the entering group. This is really a transition state effect, and it should be noted that in the five-coordinate intermediate only the *trans* ligand shares the trigonal plane with the entering and leaving group (Fig. 3.8).

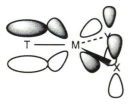

Fig. 3.8 Effect of a strong π-bonding *trans* ligand on M–T bonding

The *cis* ligands do not normally have a significant effect on reaction rate, unless they are bulky, where approach of the entering group may be inhibited. Occasionally the reaction pathway may switch from associative to dissociative.

Substitution in tetrahedral complexes

Dissociative mechanisms dominate the behaviour of tetrahedral complexes, particularly those with metals in low oxidation states, *e.g.*, metal carbonyls and tertiary phosphites such as $[Ni(CO)_4]$ and $[Ni\{P(OEt)_3\}_4]$ which obey the 18-electron rule (Section 5.1). However, associatively activated reactions are also known such as halide exchange in $[FeBr_4]^-$ (Scheme 3.7).

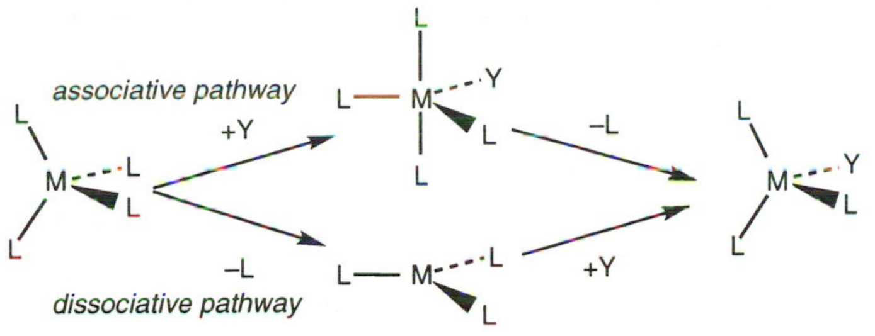

Scheme 3.7 Substitution pathways in tetrahedral complexes

An exception to the behaviour of species obeying the 18-electron rule are transition metal nitrosyls, *e.g.*, $[Co(CO)_3(NO)]$ and $[Fe(CO)_2(NO)_2]$. These compounds undergo substitution *via* an associatively activated pathway. This is possible because the NO group can bind to the metal either as a $3e^-$ or a $1e^-$ donor (Section 5.2). In most metal nitrosyl complexes, the M–N–O bond is linear ($3e^-$ donor) and the N atom is *sp* hybridised. However, the orbitals at the N atom can be rehybridised to sp^2. Here the M–N–O bond angle bends (to a maximum of 120°) and the NO becomes a $1e^-$ donor. An electron pair has been withdrawn from the metal and placed into a non-bonding orbital on N. The effect of this is to provide an empty orbital at the metal which can accommodate the entering group, thereby stabilising a five-coordinate transition state (Scheme 3.8). When the leaving group departs, the M–N–O bond straightens up. Although there is no direct structural evidence for the bending of the M–N–O bond in tetrahedral nitrosyl species, six-coordinate cobalt complexes containing such a bonding arrangement have been isolated (see Scheme 5.3).

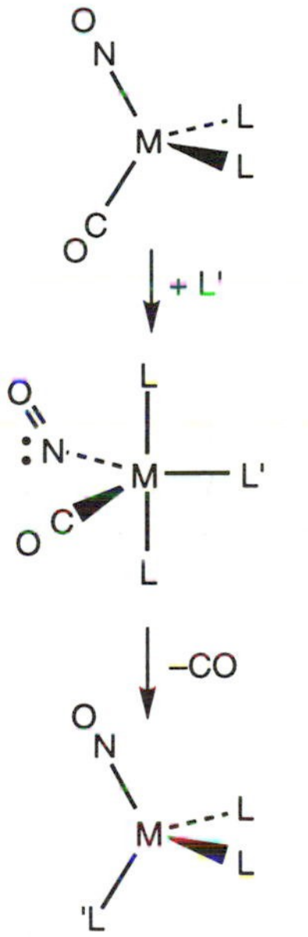

The ability of ligands to reduce electron density at the metal centre by reorganising their bonding interaction is not confined to the nitrosyl group. Substitution of CO by PR_3 in $[Mn(\eta^5\text{-}C_5H_5)(CO)_3]$ occurs *via* an associatively activated pathway. The cyclopentadienyl ligand converts from an η^5 bonding mode (for an explanation of η, the *hapto* nomenclature, see Chapter 5, p.70) to an η^3 mode. Two adjacent M–C bonds lengthen slightly, thereby effectively withdrawing an electron pair from the metal and, notionally, releasing an orbital to accommodate the entering group. Similar substitution pathways may occur in reactions involving η^3-allyl complexes, where the hydrocarbon ligand can easily rearrange to an η^1 mode, opening the way to an associative substitution process (Section 5.6).

Scheme 3.8 Substitution in tetrahedral nitrosyl complexes

3.4 Electron transfer reactions

Electron transfer reactions are often referred to as redox reactions. They may result solely in the exchange of an electron or electrons between an oxidant and a reductant, or may be accompanied by transfer of atoms, ions or molecules between the reactants. The mechanisms are of two main types: *outer sphere* in which the two reactants do not share common ligands, and *inner sphere*, in which an intermediate is formed in which there is at least one shared ligand.

Outer sphere reactions

Typical reactions would involve the reduction of $[Fe(phen)_3]^{3+}$ by $[Fe(CN)_6]^{4-}$ to $[Fe(phen)_3]^{2+}$ or oxidation of $[Fe(CN)_6]^{4-}$ by $[IrCl_6]^{2-}$ to $[Fe(CN)_6]^{3-}$. If one of the reactants is kinetically inert and does not contain a potentially bridging ligand, it is reasonable to assume an outer sphere mechanism, and that ligand exchange between the original reactants does not take place even though one of the products may be labile.

For electron transfer to occur, the distance between the two reactants must be minimised. Factors which determine the close approach of the oxidant and reductant are obviously solvent molecules and the ligands around the metal. It is necessary to be aware of the *Frank–Condon principle*: that electron rearrangements occur very much more rapidly than nuclear motion. This is because of the relative masses of the electron and nuclei. An equally important factor is that electron transfer between two reactants can only occur when the nuclei in the two participating species have positions that result in the electron having the same energy on both component. This means that the activation energy for outer sphere electron transfer is controlled largely by nuclear rearrangements necessary to attain matching energies in the oxidant and reductant.

These effects may be illustrated in the self-exchange process

$$[Fe^*(H_2O)_6]^{2+} + [Fe(H_2O)_6]^{3+} \rightleftharpoons [Fe^*(H_2O)_6]^{3+} + [Fe(H_2O)_6]^{2+}$$

$[Fe^*(H_2O)_6]^{2+}$ contains isotopically labelled iron.

The Fe–O bond lengths in the two reacting species are different: Fe–O is longer in the Fe^{II} complex than in the Fe^{III} complex because of electronic configuration and charge. A contribution to the activation energy will therefore arise because of bond length equalisation in the two reactants. If they don't become equal, then electron transfer from Fe^{II} to Fe^{III} would result in a new Fe^{II} species in which the Fe–O distances would be abnormally short: in other words one would generate an excited state species whose energy would be substantially higher than the energy required to equalise the Fe–O distances *before* electron transfer. A consequence of the inner-sphere rearrangement energy caused by bond equalisation is that the solvent shells must also be reorganised. These shells are strongly affected by electrostatic effects due to the sizes and charges of the two reacting ions. There will also be an electrostatic interaction energy since both ions are positively charged. Of course, the electron transfer reaction must be thermodynamically favourable (Section 3.2).

Self-exchange reactions proceeding by an outer sphere mechanism include the reacting pairs $[Mn(CN)_6]^{3-/4-}$, $[Fe(bipy)_3]^{3+/2+}$, $[Fe(CN)_6]^{2-/3-}$, $[Co(bipy)_3]^{3+/2+}$, $[MnO_4]^{1-/2-}$, $[Fe(H_2O)_6]^{3+/2+}$, $[Co(NH_3)_6]^{3+/2+}$, $[Co(en)_3]^{3+/2+}$ and $[Co(ox)_3]^{3-/2-}$, and reactions involving mixtures of these will also involve an outer sphere process. For example:

$$[Fe(phen)_3]^{2+} + [Fe(CN)_6]^{3-} \rightleftharpoons [Fe(phen)_3]^{3+} + [Fe(CN)_6]^{4-}$$

However, the relative rates vary enormously. For example, that for the electron exchange between $[Mn(CN)_6]^{4-}$ and $[Mn(CN)_6]^{3-}$ is greater than 10^6 L mol^{-1} sec^{-1} at 25°C, whereas that for the pair $[Fe(H_2O)_6]^{3+/2+}$ is *ca.* 4 and for $[Co(NH_3)_6]^{3+/2+}$ is *ca.* 10^{-4}. The reason for the very fast exchange rate in the manganese cyanide complexes is related to electronic configuration: low spin Mn^{II} $(t_{2g}^5) \rightarrow$ low spin Mn^{III} (t_{2g}^4). Because t_{2g} electrons have little influence on M–ligand distances unless π bonding is involved (relatively unimportant in this case), the inner-sphere rearrangement energy is quite small and consequently the solvent shell rearrangement energy is also small. X-Ray structural studies of low spin $[Fe(bipy)_3]^{2+}$ (t_{2g}^6) and $[Fe(bipy)_3]^{3+}$ (t_{2g}^5) confirm that the Fe–N bond distances do not vary much between the two oxidation states. However, in $[Fe(H_2O)_6]^{2+}$ and $[Fe(H_2O)_6]^{3+}$, the Fe–O distances are significantly different (Fe^{II}–O = 2.21 Å, Fe^{III}–O = 2.05 Å), and the rearrangement energies are consequently higher and the rate of exchange lower.

The situation with the $[Co(NH_3)_6]^{3+/2+}$ is more complicated, since the electronic configuration of Co^{II} is high-spin $[(t_{2g})^5(e_g)^2]$ whereas that of Co^{III} is low spin $[(t_{2g})^6]$. At first sight it would appear that electron transfer from a high-spin Co^{II} to a low-spin Co^{III} species would be effectively forbidden because of the differences in spin quantum number. However, it is more likely that the reason for the slow rate of exchange is due to the large difference in Co–N bond distances between high spin Co^{II} and low spin Co^{III} complexes (*ca.* 0.2 Å).

Not all Co^{II} species are high spin, however, and the electron exchange rate between the pair $[Co(bipy)_3]^{3+/2+}$ is *ca.* 9 mol^{-1} sec^{-1} at 25°C. Both of the reactants are low spin, the Co–N distances are very similar, and an electron is transferred from an e_g orbital in the reductant to an e_g orbital in the oxidant in a process perhaps facilitated by the π orbitals of the heterocyclic ligand. Hence the electron transfer rate is significantly faster than that in the ammine complex pair $[Co(NH_3)_6]^{3+/2+}$

Inner sphere reactions

There are three main steps in this process: formation of a bridged binuclear *precursor complex*, a net electron transfer from one metal centre to the other giving a *successor complex*, and dissociation of the successor complex into individual metal-containing components, with or without break up of some of these components depending on their inertness or lability. These events are illustrated in Scheme 3.9.

In this reaction, the cobalt(III) precursor $[Co(NH_3)_5X]^{2+}$ is kinetically stable whereas the chromium(II) reactant $[Cr(H_2O)_6]^{2+}$ is labile. Of the products, the chromium(III) species $[Cr(H_2O)_5X]^{2+}$ is inert but the cobalt (II)

For electron transfer from low-spin Co^{III} to high-spin Co^{II}, it would be necessary to rearrange the electronic configuration of the former to $t_{2g}^5 e_g^1$, which could be expensive in energy terms.

$[Co(NH_3)_5X]^{2+} + [Cr(H_2O)_6]^{2+}$

$\downarrow$

$\{Co^{III}(NH_3)_5–X\cdots Cr^{II}(H_2O)_5\}^{4+}$
bridged precursor

$\downarrow$

$\{Co^{II}(NH_3)_5\cdots X–Cr^{III}(H_2O)_5\}^{4+}$
bridged successor

$\downarrow$

$[Co(H_2O)_6]^{2+} + 5NH_3 +$
$[XCr(H_2O)_5]^{2+}$

Scheme 3.9 Inner sphere electron transfer for Co^{III} + Cr^{II}

product is labile. If the X group in $[Co(NH_3)_5X]^{2+}$ is specifically labelled and the reaction is carried out in the presence of high concentrations of unlabelled X in solution, it has been shown that the labelled X group is transferred essentially quantitatively to the chromium receptor, *i.e.*, the formation of a bridged intermediate is an essential step in this mechanism and intervention of X^- in the solution does not occur.

While the formation of a bridged precursor is an essential part of the inner sphere mechanism, the transfer of the bridging ligand between the partners is not. The bridged precursor can be sufficiently stable that it can be isolated, as in the reaction between $[Co(NH_3)_5(4,4'\text{-dipyridyl})]^{3+}$ and $[Fe(CN)_5(H_2O)]^{3-}$ shown in Fig. 3.9.

It is also possible to isolate bridged dinuclear successor complexes, as in the reaction between $[Fe^{III}(CN)_6]^{3-}$ and $[Co^{II}(CN)_5(H_2O)]^{3-}$ when $[(NC)_5Fe\text{–}CN\text{–}Co(CN)_5]^{6-}$ is formed.

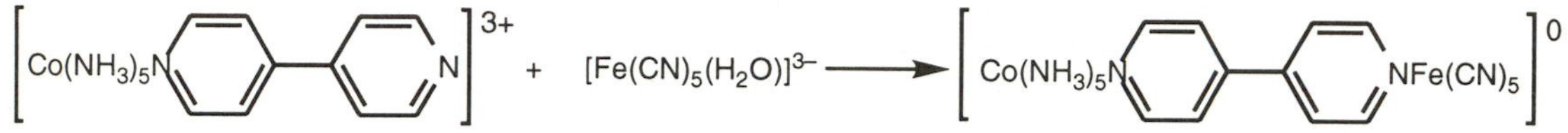

$$\left[Co(NH_3)_5N\!\!\!\bigcirc\!\!\!\bigcirc\!\!\!N \right]^{3+} + [Fe(CN)_5(H_2O)]^{3-} \longrightarrow \left[Co(NH_3)_5N\!\!\!\bigcirc\!\!\!\bigcirc\!\!\!NFe(CN)_5 \right]^{0}$$

Fig. 3.9 Isolable precursor complex in inner sphere electron transfer

For an inner sphere process, the two participating metal centres can be separated by several atoms and electron transfer can still occur at a significant rate. This can obviously be facilitated when the bridging ligand is delocalised, as in the 4,4'-dipyridyl shown in Fig. 3.9. Aromatic and other unsaturated ligands can readily conduct electrons.

A similar situation can occur in the reaction between the N-bonded thiocyanato complex $[Fe(H_2O)_5(NCS)]^{2+}$ and $[Cr(H_2O)_6]^{2+}$, when the S-bonded thiocyanato product, $[Cr(H_2O)_5(SCN)]^{2+}$ is formed. This is the kinetically stable product which isomerises to the thermodynamically more stable $[Cr(H_2O)_5(NCS)]^{2+}$ (Cr^{II} is relatively hard, N is hard and S is soft). It is presumed that the bridged precursor is $[Fe(H_2O)_5\text{–}NCS\text{–}Cr(H_2O)_5]^{4+}$ in which the two metals and the thiocyanato group form a chain. However, in a similar reaction between $[Co(NH_3)_5(N_3)]^{2+}$ and $[Cr(H_2O)_6]^{2+}$ it is evident that the chromium binds *via* only one nitrogen atom adjacent to the cobalt (Fig. 3.10).

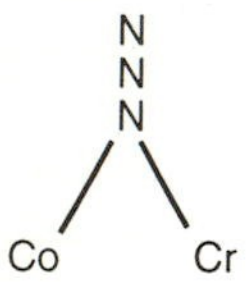

Fig. 3.10 Azido-bridged intermediate

4 Compounds in higher oxidation states

As mentioned in Sections 1.1 and 3.2 the thermodynamic stability of the highest oxidations states gradually decreases across the transition series from Ti^{IV} to Fe^{VIII}, the last two (manganese and iron) being particularly unstable and, therefore, powerfully oxidising.

In contrast, oxidation state III progressively increases in stability, although Cr^{III} is an early exception because it is kinetically inert (d^3). Cobalt(III), with a spin-paired d^6 configuration, is also kinetically stable, and many reactions can be performed on the ligands which cannot be easily achieved, if at all, in absence of the metal.

This chapter is divided into several sections, dealing with oxo, nitrido, imido and amido species, particularly those containing M=O and M≡N bonds, some organometallic species, reactions of coordinated ligands, and metal complexes where the ligand rather than the metal is oxidised.

4.1 Oxo, aqua, amino and related compounds

From Ti to Mn, the highest oxidation states can only be stabilised by oxo, fluoro or chloro ligands. No fully aquated ions exist for these elements in oxidation states higher than IV, partly because of the high polarisability of the metal ion and partly because of the ability of the oxo ligand to π-donate to the metal (Fig. 4.1).

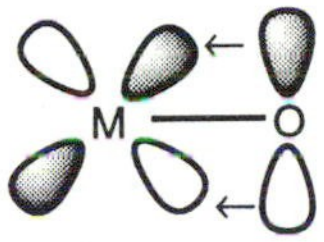

Fig. 4.1 O $2_{p\pi} \rightarrow M_{d\pi}$ donation

Oxoanions and isopolyanions

The most commonly encountered simple oxoanions are the tetrahedral $[MnO_4]^-$ and $[CrO_4]^{2-}$, both of which are powerfully oxidising (Section 3.2). Other oxoanions do exist, but rarely in simple form in water: they usually occur in solid mixed-metal oxides and minerals. Typical of these are the metatitanates, formally containing either $[Ti^{IV}O_3]^{2-}$, as in ilmenite ($FeTiO_3$) and perovskite ($CaTiO_3$), or orthotitanates, formally containing $[Ti^{IV}O_4]^{4-}$, as in Ba_2TiO_4. The vanadate ion, $[V^VO_4]^{3-}$, is stable at very low pH, but oligomerises to more complex chain or cluster structures as the pH decreases (see below).

The M–O bond lengths in oxoanions is 1.5–1.8 Å, depending on coordination number. The bond order is usually slightly greater than two since there are two p_π orbitals on the oxygen atom and there is extensive $p_\pi \rightarrow d_\pi$ donation (Fig. 4.1).

The behaviour of oxoanions as a function of pH is very complex. Although aquated Ti^{4+} salts do not exist, the ion may exist in strong acid solution, perhaps as $[Ti(OH)_2]^{2+}$ or $[Ti(OH)Cl]^{2+}$, but as the pH is increased, complex ions, *e.g.*, $[Ti_3O_4]^{4+}$, are formed at first and these eventually lead to colloidal or precipitated hydrous titanium oxide, $TiO_2.nH_2O$. Vanadium pentoxide dissolves in sodium hydroxide giving colourless $[VO_4]^{3-}$, stable at pH > 13. As the pH is reduced, a protonated species, $[VO_3(OH)]^{2-}$ is formed first and then aggregation occurs, leading ultimately to a series of orange decavanadates and finally aquated $[VO_2]^+$:

$$[VO_4]^{3-} + H_2O \rightleftharpoons [VO_3(OH)]^{2-} + OH^-$$

$$2[VO_3(OH)]^{2-} \rightleftharpoons [V_2O_7]^{4-} + H_2O$$

$$[VO_3(OH)]^{2-} + H^+ \rightleftharpoons [VO_2(OH)_2]^-$$

$$3[VO_2(OH)_2]^- \rightleftharpoons [V_3O_9]^{3-} + 3H_2O$$

$$4[VO_2(OH)_2]^- \rightleftharpoons [V_4O_{12}]^{4-} + 4H_2O$$

$$10[V_3O_9]^{3-} + 15H^+ \rightleftharpoons 3[V_{10}O_{27}(OH)]^{5-} + 6H_2O$$

$$[V_{10}O_{27}(OH)]^{5-} + H^+ \rightleftharpoons [V_{10}O_{26}(OH)_2]^{4-}$$

$$[V_{10}O_{26}(OH)_2]^{4-} + 14H^+ \rightleftharpoons 10[VO_2]^+ + 8H_2O$$

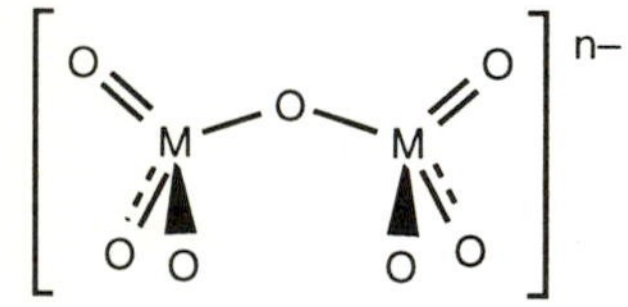

Fig. 4.2 Structure of $[M_2O_7]^{n-}$ (M = V, Cr)

Monomeric vanadium oxoanions are formed only in very dilute solutions. While $[VO_4]^{3-}$ is a simple tetrahedron, $[V_2O_7]^{4-}$ is made up of two VO_4 tetrahedra sharing a corner (Fig. 4.2). The decavanadate ions are comprised of 10 fused $\{VO_6\}$ octahedra. Although these species all contain V^V, isopolyvanadates containing V^{IV} can be obtained by dissolving $[VO(H_2O)_5]^{2+}$ (which contains a V=O bond) in alkali, which initially gives a grey hydrated oxide $VO_2.nH_2O$ similar to Ti^{IV}, but this subsequently dissolves to afford brown solutions containing $[V_{18}O_{42}]^{12-}$ constructed of fused pyramidal $\{VO_5\}$ groups. The VO_2^+ group has a *cis* rather than a *trans* arrangement as this favours better $O_{p\pi} \rightarrow V_{d\pi}$ bonding.

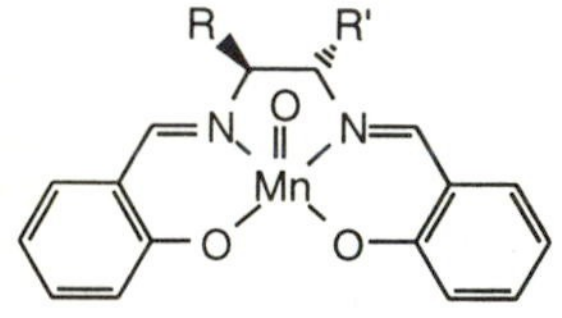

Yellow $[CrO_4]^{2-}$ is stable in aqueous media with pH > 6 and is converted to the orange–red dichromate, $[Cr_2O_7]^{2-}$ (Fig. 4.2) on acidification. Chromate ion is smaller than vanadate, and is apparently limited to tetrahedral coordination, so the formation of isopolyanions, which require $\{CrO_6\}$ octahedra, is not observed.

The paramagnetic manganate ion, $[Mn^{VI}O_4]^{2-}$ (d^1) is stable only in very basic solution, and readily disproportionates into diamagnetic permanganate, $[Mn^{VII}O_4]^-$ and $Mn^{IV}O_2$. Ferrate ion, $[Fe^{VI}O_4]^{2-}$, is extremely unstable in solution, but is isostructural with $[Mn^{VI}O_4]^{2-}$ and is a stronger oxidant than $[Mn^{VII}O_4]^-$. Comparable simple ions of Co and Ni do not appear to exist although tetrahedral $[Co^{IV}O_4]^{4-}$ occurs in solid Na_4CoO_4.

The oxymanganese (IV) salen complex shown below, obtained by reaction of [MnCl(salen)] with NaOCl, catalyses the epoxidation of various classes of alkenes.

Aqua ions

The $3d$ transition metals all form hexa-aqua metal ions, $[M(H_2O)_6]^{n+}$ in at least one oxidation state. Only Ti does not form $[M(H_2O)_6]^{2+}$ and only Ni and Cu do not form $[M(H_2O)_6]^{3+}$. These ions are normally octahedral, although the ions containing Cr^{2+}, Mn^{3+} and Cu^{2+} are distorted because of the Jahn–Teller effect (Section 1.4, p. 16).

While most of the hexa-aqua ions are labile, $[Cr(H_2O)_6]^{3+}$ is stable because of the electronic configuration of Cr^{3+} $(t_{2g})^3$ (Section 3.1). However, the aqua ions are more or less acidic, dissociating to form aqua-hydroxo species:

$$[M(H_2O)_6]^{n+} \rightleftharpoons [M(H_2O)_5(OH)]^{(n-1)+} + H^+$$

The pK_a for $[Cr(H_2O)_6]^{3+}$ is *ca.* 4 and for $[Fe(H_2O)_6]^{3+}$ just over 3. This means that $[Cr(H_2O)_6]^{3+}$ is almost as strong an acid as formic acid, and the consequences for aqueous Cr^{III} chemistry are important, as shown in Scheme 4.1.

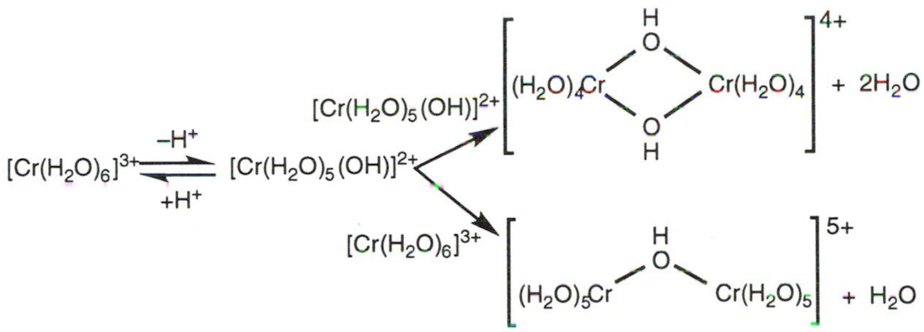

Scheme 4.1 Hydrolysis or acid dissociation of $[Cr(H_2O)_6]^{3+}$

Ammine complexes cannot be prepared by addition of NH_3 to $[Cr(H_2O)_6]^{3+}$ in aqueous media, partly because of the very high kinetic stability of Cr^{III} in this environment, and also because of deprotonation of H_2O.

Similar acid dissociation behaviour occurs with $[Fe(H_2O)_6]^{3+}$ which is stable at pH < 1. Just above this pH, this ion exists in equilibrium with $[Fe(H_2O)_5(OH)]^{2+}$ and perhaps with $[Fe(H_2O)_4(OH)_2]^+$. However, $[Fe(H_2O)_5(OH)]^{2+}$ exists in equilibrium with $[(H_2O)_5Fe-O-Fe(H_2O)_5]^{4+}$. As the pH increases, more aggregated species are formed until colloidal gels and finally a red–brown gelatinous hydrous oxide are produced.

$[Co(H_2O)_6]^{3+}$ is unstable in aqueous solution (Section 3.2), oxidising H_2O to O_2 and forming $[Co(H_2O)_6]^{2+}$.

Oxo anion as a bridging ligand

The oxo anion can bridge two, three or four metal atoms. In the first mode, the M–O–M bond angle can be linear or bent. The linear arrangement is relatively rare with first-row transition metals, but does occur in $[(NH_3)_5Cr-O-Cr(NH_3)_5]^{4+}$, prepared by deprotonation of $[(NH_3)_5Cr(\mu-OH)Cr(NH_3)_5]^{5+}$, many $\{Fe-O-Fe\}^{4+}$ complexes and in $[(tpp)FeOFe(tpp)]$ {tpp =

Rusting of iron involves formation of the hydrated species $Fe(OH)_3$ or $FeO(OH)$, and is an electrochemical process (see Section 3.2) which involves water, O_2 and an electrolyte. The latter can consist of iron(II) sulphates, formed from atmospheric SO_2, or Cl^- arising from sea spray or from treatment of ice on roads. Protection against rust can be achieved by galvanising with Zn which acts as a sacrificial anode, the Zn being attacked before Fe because it is higher in the electrochemical series. Iron surfaces may also be treated with inhibitors such as CrO_4^{2-} or phosphate, all of which produce a protective coating of Fe_2O_3.

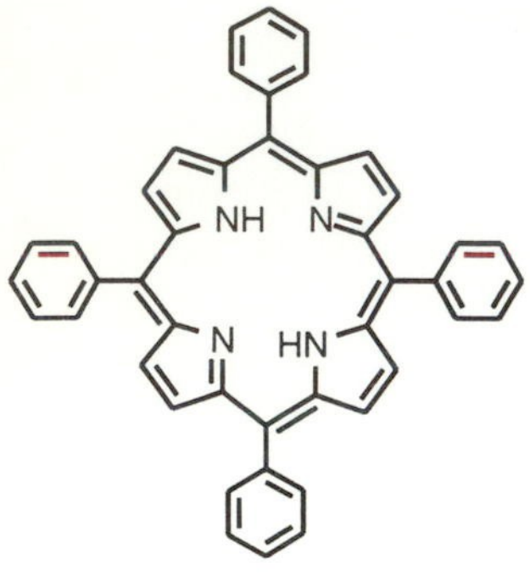

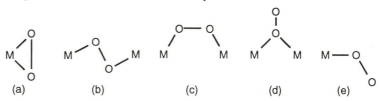

Tpp, tetraphenylporphyrin

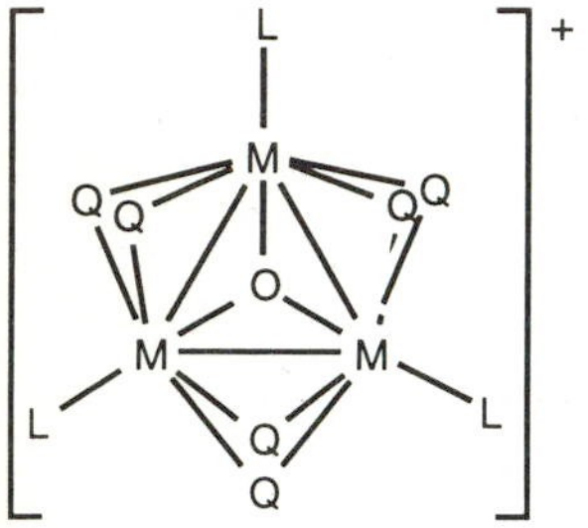

Fig. 4.3 Structure of $[M_3O(O_2CR)_6L_3]^+$, Q = bridging acetate group, OAc^-

tetraphenylporphyrinate(2–)}. The relative shortness of the Cr–O bond distances and the linearity of the Cr–O–Cr group indicate significant $Cr_{d\pi}$–$O_{p\pi}$ interaction.

Bent M–O–M bond angles occur in $[M_2O_7]^{n-}$ (M = V, n = 4; M = Cr, n = 2) and in $[Cl_3Fe–O–FeCl_3]^{2-}$ although a linear arrangement can be induced in the latter when crystallised with a large cation.

Three metal atoms can be bridged by O^{2-} in the 'basic' carboxylates, $[M_3O(O_2CR)_6L_3]^+$ (M = V–Co; L = H_2O, py; Fig. 4.3). The M_3O core is essentially triangular and planar because of significant M–O π bonding. The monocation contains M^{III}, but the species may be oxidised or reduced in a one-electron step giving rise to mixed-valence trimetallic cores, *e.g.*, $[Mn_3O(O_2CMe)_6(py)_3]$. This species is formally described as containing $Mn^{II}Mn^{III}Mn^{III}$ although the spectroscopic and magnetic properties indicate that the complex has a delocalised ground state with formal metal oxidations states of 2.66. Mixed metal 'basic' carboxylates, *e.g.*, $[Cr_2MnO(O_2CMe)_6(py)_3]$ and $[CrFe_2O(gly)(H_2O)_3]^{7+}$ (gly = glycinate ion), have structures similar to $[M_3O(O_2CR)_6L_3]^+$. The mineral metavoltine contains $[Fe_3O(SO_4)_6(H_2O)_3]^{5-}$.

The O atom in $[Fe_4O(O_2CMe)_{10}]$ bridges all four Fe^{III} centres.

Dioxygen, superoxo and peroxo compounds

The O_2 molecule and its mono-reduced, O_2^- (superoxo), and di-reduced, O_2^{2-} (peroxo) forms act as ligands to first-row transition metals. Molecular oxygen reacts reversibly with some metal complexes and such reactions are involved in the transport and storage of dioxygen by haemoglobin and other metalloproteins, and are described in Chapter 6.

Fig. 4.4 Modes of bonding of O_2 with metals

In reactions of complexes with O_2 the metal is normally oxidised and the dioxygen reduced, the mode of bonding (Fig. 4.4) of the O_2 moiety depending partly on the extent of electron transfer. Monodentate, bent O_2^- complexes (Fig. 4.4(e)) occur in $[Co(O_2)(CN)_5]^{3-}$. Bridging superoxo species (Fig. 4.4.(c)) also occur in cobalt(III) chemistry, for example, following one-electron oxidation of the peroxo species $[(NH_3)_4Co(\mu\text{-}O_2)(\mu\text{-}NH_2)Co(NH_3)_4]^{3+}$. The superoxo ligand distances are in the range 1.10–1.30 Å, which may be compared with 1.33 Å in simple superoxide salts.

Peroxo complexes may either be prepared by direct reaction of a complex with O_2, or by reaction with hydrogen peroxide:

$$[Co^I(diars)_2]^+ + O_2 \rightarrow [Co^{III}(O_2)(diars)_2)]^+$$

$$[CrO_3(OH)]^- + 2H_2O_2 + H^+ \rightarrow [CrO(O_2)_2(H_2O)] + 2H_2O$$

In these species the dioxygen moiety is sideways bonded (Fig. 4.4(a)), the O–O distances in the range 1.40–1.50 Å (compared with 1.49 Å in peroxide salts). The bonding is similar to that in metal alkene complexes (Section 5.4), in that σ bonding involves donation from an O π orbital to an acceptor orbital on the metal, and back donation from a metal $d\pi$ orbital to the π^* orbitals of O_2. Chromium forms a variety of peroxo complexes, *e.g.*, $[Cr(NH_3)_3(O_2)_2]$ which contains Cr^{IV} and has a pentagonal bipyramidal structure (axial NH_3), $[Cr(O_2)_4]^{3-}$ which contains Cr^V with a dodecahedral structure, and $[CrO(O_2)_2(py)]$ which contains Cr^{VI} and has a pentagonal pyramidal structure (axial Cr=O).

Alkoxo and aryloxo complexes

Just as metal ions dissolve in water forming hexa-aqua species, in alcohols, particularly methanol, similar species such as $[M(ROH)_6]^{n+}$ are formed. The alcohol is readily displaced by stronger donors, including water.

Transition metal alkoxides are normally prepared by reaction of metal halides MX_n with appropriate alcohols in the presence of an HX acceptor. They may also be obtained by metathesis of metal halides with alkali metal or Tl^I alkoxides:

$$TiX_4 + 4ROH + 4NEt_3 \rightarrow [Ti(OR)_4] + 4\,[NHEt_3]X$$

These alkoxides are readily hydrolysed but are thermally stable, distillable liquids or volatile solids.

Metal alkoxides are extremely reactive, undergoing a wide range of 'insertion' reactions, for example, with CO_2 to give carbonate esters $\{M(O_2COR)\}$ and with CS_2 to give xanthates $\{M(S_2COR)\}$.

The majority of simple alkoxide complexes are oligomeric, containing μ_2- and μ_3-OR groups, *e.g.*, $[\{Ti(OEt)_4\}_4]$, Fig. 4.5, $[\{Mn(OCHBu^t)_2\}_3]$, which contains a $\{OMn(\mu\text{-}O)_2Mn(\mu\text{-}O)_2MnO\}$ core, and $[\{Cu(OBu^t)\}_4]$. However, when the alkoxide is very bulky, *e.g.*, $OCBu^t_3$, $OSiBu^t_3$, $2,6\text{-}OC_6H_3(Bu^t)_2$, adamantoxide, monomeric species can be formed, *e.g.*, $[Ti(OC_6H_3Bu^t_2)_3]$, distorted paramagnetic tetrahedral $[Cr^{III}(OCHBu^t_2)_4]^-$ and $[Cr^{IV}(OBu^t)_4]$.

The M–OR bond is often shorter than expected for a bond order of unity, caused partly by $p_\pi \rightarrow d_\pi$ donation from O to the metal.

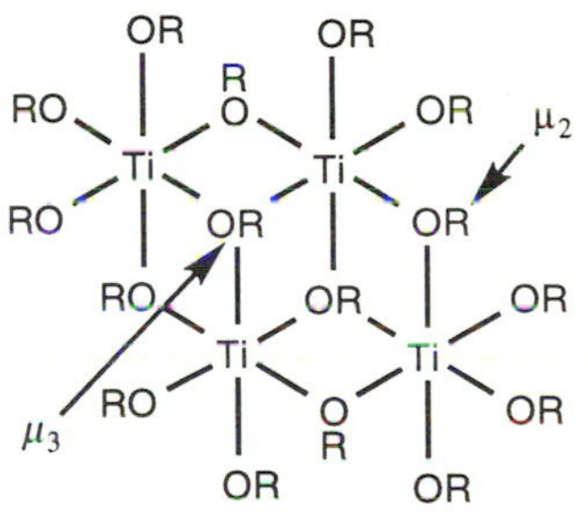

Amido, alkylamido, imido and nitrido complexes

The amido group, NH_2^-, and its homologues, the alkylamides, NHR^- and NR_2^-, are isoelectronic with OH^- and OR^-. The amido group is implicated in base-catalysed hydrolysis of cobalt(III) ammine complexes (Section 3.3) and, like O^{2-}, frequently occurs as a bridging group, as in $[(NH_3)_4Co(\mu\text{-}O_2)(\mu\text{-}NH_2)Co(NH_3)_4]^{3+,4+}$.

Metal complexes of mono- and di-alkylamides are generally made by reaction of appropriate halides with LiNHR or LiNR_2. Like their alkoxo analogues, the compounds are either volatile liquids or solids, and are very sensitive to hydrolysis by water. They are readily converted to alkoxides by reaction with ROH. Also like their alkoxo analogues, they undergo 'insertion' reactions with CO_2, giving carbamates, *e.g.*, $[Ti(O_2CNR_2)_4]$ and with CS_2 giving dithiocarbamates, $[Ti(S_2CNR_2)_4]$ (Section 3.3, p.41).

Because of the ease of hydrolysis of titanium alkoxides to TiO_2, these materials are used in paints to confer scratch resistance, waterproofing and heat resistance. They are also used in combination with β-diketones, which give $[Ti(OR)_2(\beta\text{-diketonate})_2]$, to make *thixotropic* or drip-and-run-resistant paints.

Fig. 4.5 Structure of $[\{Ti(OR)_4\}_4]$

Metal alkylperoxo complexes are implicated in the epoxidation of alkenes. Mixtures of $[\{Ti(OR)_4\}_n]$ with tartrate esters and ROOH give species such as $[Ti_2(OR)_2(dipt)]$ (dipt = di-isopropyltartrate) which expoxidise allylic alcohols (the *Sharpless* reaction).

Complexes containing NHR may readily eliminate H, affording bridging alkylimido groups, μ-NR:

$$2[Ti(NHR)_4] + 4NHR'_2 \rightarrow [(R'_2N)_2Ti(\mu\text{-}NR)_2Ti(NR'_2)_2] + 6NH_2R$$

Compounds containing bulky NR_2 groups are mostly monomeric, as in $[M(NR_2)_3]$ (R = Et, M = Cr, Fe; R = SiMe₃, M = Ti, V, Fe) and $[Ni(NPh_2)_3]^-$. There is evidence for significant $Np_\pi \rightarrow Md_\pi$ donation and this may account for the trigonal planar geometry of the $M(NR_2)_3$ group. The NR_2 groups can also act as bridges, as in $[(R_2N)M(\mu\text{-}NR_2)_2M(NR_2)]$ (M = Co, Ni) in which the metal atoms are also trigonal planar.

Alkylimido groups, NR^-, can be formed by oxidation of primary amines, by eliminating either CO_2 in reactions of oxo-metal species with organocyanates, or N_2 in reactions of low-oxidation state species with organoazides:

$$CrO_3 + 2NH(SiMe_3)Bu^t \rightarrow [Cr^{VI}(NBu^t)_2(OSiMe_3)_2] + H_2O$$

$$VOCl_3 + ArNCO \rightarrow [V^V(NAr)Cl_3] + CO_2$$

$$[V(\eta^5\text{-}C_5Me_5)_2] + PhN_3 \rightarrow [V^{IV}(\eta^5\text{-}C_5Me_5)_2(NPh)] + N_2$$

The M–NR group may be linear or bent (Fig. 4.6), acting as a *4e* donor in the former arrangement because of $N_{p\pi} \rightarrow M_{d\pi}$ donation, and as a two electron donor in the latter. This situation is comparable to that in NO complexes (Section 5.2).

The nitrido ligand, N^{3-}, is rare in first-row transition metal chemistry. It is present in $[Mn^VN(tpp)]$ (for tpp see p.50) and acts as a bridging group in $[(tpp)FeNFe(tpp)]$. This compound has one unpaired electron, and is a mixed valence compound formally containing Fe^{III} and Fe^{IV} but there is significant delocalisation in the Fe–N–Fe group (oxidation state averages 3.5). It may be oxidised and reduced in one-electron steps to a monocation (containing two Fe^{IV}) and a monoanion (2 Fe^{III}) and may be compared to $[(tpp)FeOFe(tpp)]$ which contains two Fe^{III} centres.

> $[Mn(NBu^t)_4]^-$ is isoelectronic with $[MnO_4]^-$ and is obtained by reaction of $[Mn(NBu^t)_3Cl]$ with $LiNHBu^t$.

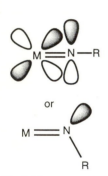

Fig 4.6 Bonding in M=NR group

4.2 Organometallic compounds

A small number of important organometallic compounds of the first-row transition metals are known in oxidation states III, IV and, occasionally, V, and titanium organometallics are extremely important in polymerisation catalysis. Most of these compounds are prepared by reaction of appropriate metal halides with Grignard or alkyl lithium reagents, or with aluminium alkyls. For some metals, particularly Cr–Co, oxidation or disproportionation accompanies formation of the alkyl compound, affording species in relatively high oxidation states

Simple organometallic compounds

Simple metal alkyls, such as $FeEt_2$ or $CoPr_3$ are extremely labile, undergoing decomposition pathways involving hydrogen transfer. This instability is due to kinetic rather than to thermodynamic factors, any low-coordinate metal alkyl in which there is a β-hydrogen atom readily eliminating alkene with the formation of potentially unstable hydrido-metal derivatives (Fig. 4.7). This process can be stopped by designing molecules in which decomposition pathways are blocked: either (a) by avoiding alkyl groups containing β-hydrogen atoms or (b) by ensuring the absence of empty coordination sites at the metal.

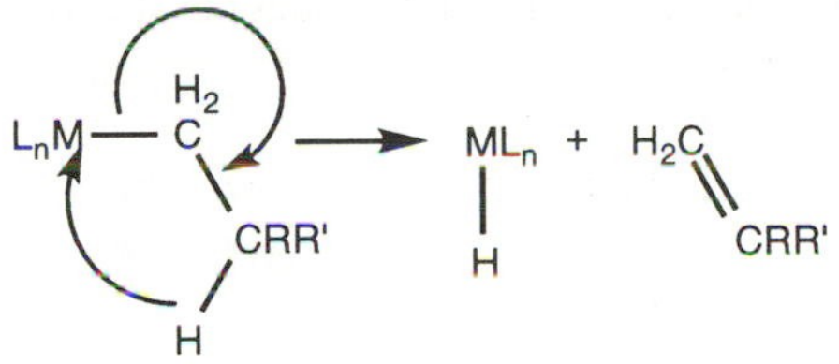

Fig. 4.7 β-elimination pathway for metal alkyls

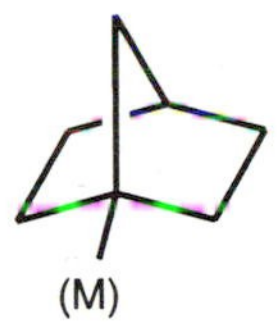

Alkyl groups which contain no β-hydrogen atoms include CH_2Ph, CH_2SiMe_3, $CH(SiMe_3)_2$, cyclohexyl and 1-norbornyl (nor) (Fig. 4.8). Examples of the β-elimination-stabilised metal alkyl derivatives include tetrahedral $[M(CH_2SiMe_3)_4]$ and trigonal $[M\{CH(SiMe_3)_2\}_3]$ (M = Ti, V), the ligands in the latter group being too bulky to permit four-coordination. The 1-norbornyl complexes $[M(nor)_4]$ where M = Ti–Co, are the remarkably stable. They probably have a distorted tetrahedral structures and, with the exception of $[Ti(nor)_4]$, are paramagnetic. $[Co(1\text{-norbornyl})_4]$ has a magnetic moment of 2.0 BM, suggesting the presence of one unpaired electron. If this complex is truly tetrahedral, then the alkyl ligand is exerting an extremely strong ligand field since an overwhelming majority of tetrahedral d^5 species are high spin in this geometry (Section 1.4). It is possible that the molecule is significantly structurally distorted, however. It is obtained by reaction of simple cobalt salts with Li(nor) in an ether solvent, which gives the readily oxidised paramagnetic cobaltate(II) anion $[Co(nor)_4]^{2-}$. The $Co(nor)_4$ core is so stable that it may be reduced to paramagnetic $[Co(nor)_4]^-$ (μ = 3.18 BM) and oxidised to diamagnetic $[Co(nor)_4]^+$.

Adopting the alternative strategy (b), unstable alkyls such as $TiMe_4$ (decomposition –40°C) and $MnMe_4$ can be stabilised by coordination with a bidentate ligand, as in the six-coordinate $[TiEt_4(bipy)]$ and $[MnMe_4(dmpe)]$ (dmpe = $Me_2PCH_2CH_2PMe_2$). The cobalt dimethylglyoximato complexes $[Co(CH_2CH_2R)(dmgH)_2L]$ (L = Lewis base, Fig. 6.14(b), p.89) are stabilised similarly. A very large numbers of metal carbonyl alkyl derivatives, *e.g.*, $[Mn(CO)_5Et]$, $[Fe(\eta\text{-}C_5H_5)(CO)(PPh_3)Et]$, and their analogues also fall into this category, although the oxidation state of the metal is not especially high.

(M)

Fig. 4.8 1-Norbornyl group, nor

Alkene polymerisation (Ziegler–Natta) catalysis

Mixtures of $TiCl_4$ and $AlEt_3$ in hydrocarbon solvents heterogeneously catalyse the polymerisation of ethene to high-density polyethylene (a relatively high-melting linear polymer). This reaction is different to the radical polymerisation of ethene under high pressure which gives a low-density, relatively low-melting branched polymer. The titanium chloride/aluminium ethyl mixture reacts to give a finely divided form of solid

TiCl$_3$ in which the metal atom is octahedrally surrounded by Cl atoms, and there are ethyl groups dispersed on the surface. Catalysis is thought to occur at coordinatively vacant sites on the surface adjacent to the alkyl groups, and the surface structure is believed to be responsible for the stereoregular polymerisation. For example, propene can be produced in the highly stereoregular isotactic form (Fig. 4.9) which has high density, hardness and tensile strength. A possible mechanism for this process is shown in Scheme 4.2.

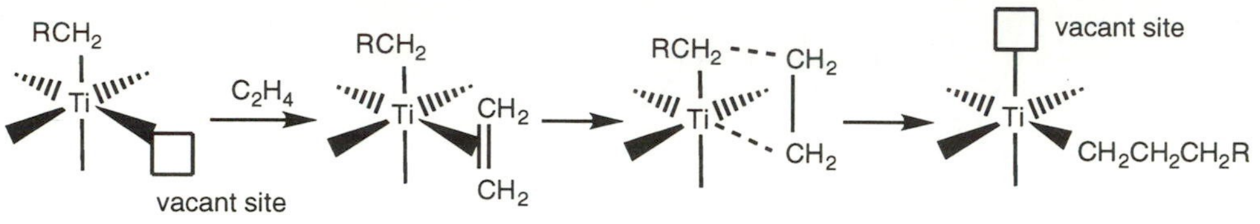

Scheme 4.2 Ziegler–Natta alkene polymerisation

The key points of the above mechanism is the involvement of a vacant coordination site at the TiCl$_3$ surface, adjacent to a Ti–CH$_2$R group, accommodating an alkene molecule. Group migration of the CH$_2$R group to the alkene can then occur leading to chain propagation. Chain termination occurs *via* β-hydrogen elimination from an elongated alkyl chain, leading to the formation of a titanium hydrido surface species which, in turn, reacts with fresh alkene regenerating the Ti–CH$_2$R surface species. Support for the group migration process has been obtained from the observation that [Co(η^5-C$_5$H$_5$)(η^2-C$_2$H$_4$)(CD$_3$)$_2$] can be interconverted to [Co(η^5-C$_5$H$_5$)(CH$_2$CH$_2$CD$_3$)(CD$_3$)].

Bis(cyclopentadienyl)titanium species, generated by reaction of [Ti(η-C$_5$H$_5$)$_2$Cl$_2$] with a Lewis acid activator, can give highly active homogeneous Zeigler–Natta catalysts, affording polymers of high stereoregularity. The active species is thought to be [Ti(η-C$_5$H$_5$)$_2$R]$^+$, possibly solvated, but which readily binds an alkene giving [Ti(η-C$_5$H$_5$)$_2$R(η^2-CH$_2$=CHR]$^+$. The mechanism of polymerisation is very similar to that in Scheme 4.2.

4.3 Reactions at coordinated ligands

The majority of reactions described in this chapter are concerned with exchange of ligands at metals. However, coordination of ligands to metals can profoundly influence the reactivity of those ligands to other reagents. This can happen in two ways: (a) by attaching the ligand to the metal which acts as a Lewis acid or acceptor, polarising electron density within the ligand towards the metal, or (b) by π back-donation from the metal to the ligand, which makes the latter relatively 'electron rich'. The consequences of the latter for subsequent ligand reactivity are described in Chapter 5 (p. 64, 66).

When the metal acts as a Lewis acid, it is normally seen as a 'hard' acid, and this is typical for first-row transition metals in high oxidation states. Coordination of certain types of ligands to such 'hard' acids tends to make them more susceptible to nucleophilic attack, and many reactions of kinetically stable Cr^{III} and Co^{III} complexes demonstrate this behaviour.

The hydrolysis of peptides, esters and amides by hydroxide ion is catalysed by 'hard' metal ions such as Co^{2+} and Cu^{2+}, but the mechanism of reaction

For an explanation of the η nomenclature, see Section 5.6, p.70.

isotactic
all C atoms have the same
configuration

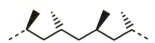

syndiotactic
regular change of
configuration

atactic
irregular change of
configuration

Fig. 4.9 Polymer structure

is best illustrated by utilising amine complexes of Co^{3+}. The hydrolysis of a glycine ester by a cobalt tetramine complex is illustrated in Scheme 4.3.

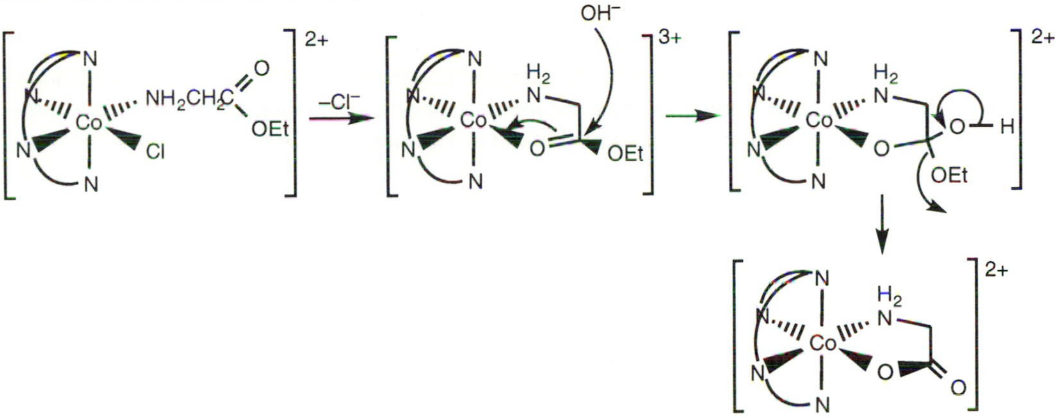

Scheme 4.3 Hydrolysis of glycine esters assisted by cobalt(III) complexes

The formation of a five-membered chelate ring in the intermediate species is thermodynamically favoured, and coordination of the carbonyl group polarises electron density in that region of the ligand, making it more susceptible to nucleophilic attack by OH^-.

Occasionally, the hydroxide ion is coordinated by the metal which places it in the correct position to cause efficient hydrolysis. This is shown in Scheme 4.4, in which the rate of hydrolysis of the nitrile is 10^{11} times faster compared with the free base.

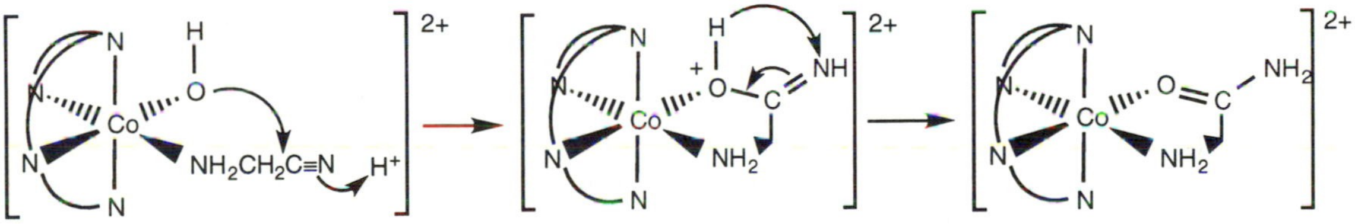

Scheme 4.4 Hydrolysis of a non-bonded nitrile group assisted by cobalt(III) complexes

Metal complexes of β-diketonates, particularly acetylacetonates (pentane-2,4-dionates) are very stable and exhibit properties which are typical of aromatic systems. Halogenation at the 4-position on the chelate ring is achieved using *N*-halosuccinamides, and the complexes may be further derivatised by substitution at the C–X bonds (Scheme 4.5).

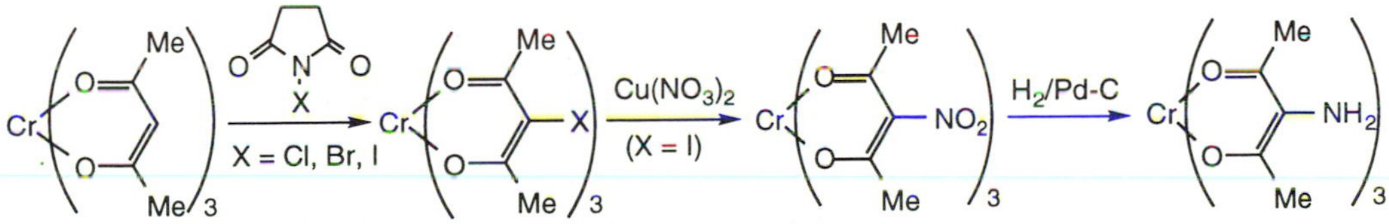

Scheme 4.5 Reactions at coordinated β-diketonato chromium(III) complexes

Metal ions can be very effective in assisting in the formation of macrocyclic rings which are otherwise very difficult or impossible to prepare.

The size of the metal ion can be important in that the cation serves to hold a partially formed ligand in position so that the remainder of the ligand can be assembled. This is the *template* effect, and many complexes of Fe^{II}, Co^{II}, Ni^{II} and Cu^{II} are formed by exploitation of this effect. However, a macrocycle can be designed to generate a 3-D cage about the metal, a process known as *encapsulation*. An example of such a reaction is the treatment of the tris(dimethylglyoximato) complex of Fe^{III} or Co^{III} with $BF_3.OEt_2$ which results in the formation of a *clathro-chelate* or *cryptate* complex, *e.g.*, $[Co\{FB[ONC(Me)C(Me)CNO]_3BF\}]^+$, Fig. 4.10(a).

A similar type of encapsulation process occurs when $[Co(en)_3]^{3+}$ reacts with formaldehyde and ammonia giving the so-called *sepulchrate* species shown in Fig. 4.10(b). It is not possible to preform the ligand, and, even if it was, the metal ion could not be inserted. Equally, removal of the metal ion is extremely difficult without first destroying the ligand.

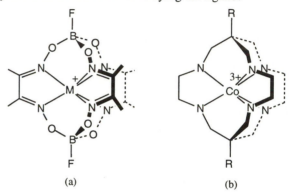

(a)   (b)

Fig. 4.10 (a) A metal *cryptate*; (b) a cobalt *sepulchrate*

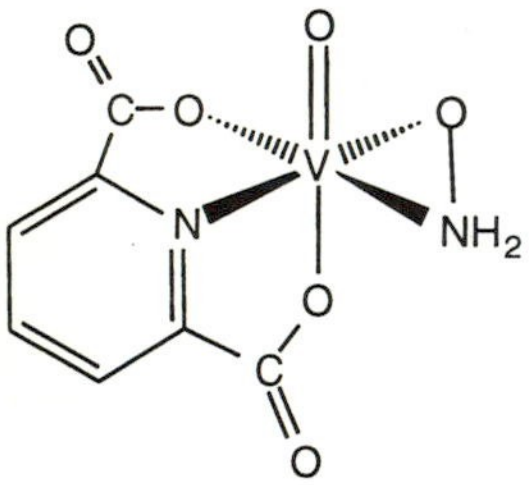

Fig. 4.11 $[V(O)(\eta^2\text{-}NH_2O)$-$(dipic)(H_2O)]$

Unstable intermediates in conversion of one species to another can occasionally be intercepted by coordination to a metal. An example is the progressive oxidation of hydroxylamine to NO. Hydroxylamines can react with vanadium(V) species giving monodeprotonated hydroxylaminates, *e.g.*, $[(\eta^2\text{-}Et_2NO)_2V(O)\text{–}O\text{–}V(O)(\eta^2\text{-}Et_2NO)_2]$, where the Et_2NO^- ligand is bidentate, forming VON three-membered rings. However, when vanadium(V) species react with hydroxylamine itself, similar species can only be trapped by addition of the dipicolinate ligand, giving $[V(O)(\eta^2\text{-}NH_2O)(dipic)(H_2O)]$ (Fig. 4.11). Further deprotonation of this affords first a η^2-NHO species and finally a nitrosyl complex (Scheme 4.6; Section 5.2).

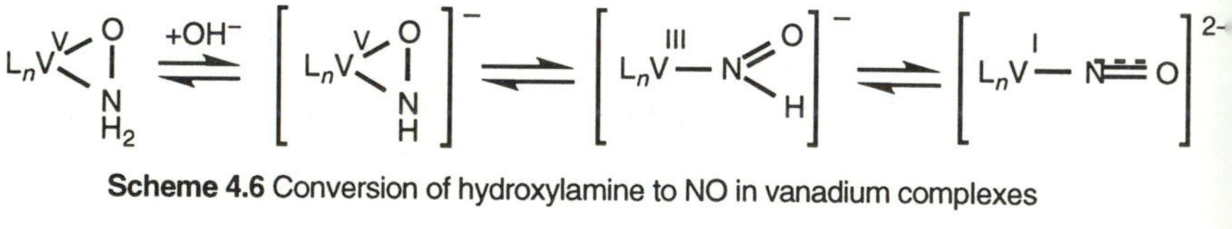

Scheme 4.6 Conversion of hydroxylamine to NO in vanadium complexes

5 Compounds in lower oxidation states

For the earlier transition metals, particularly Ti and V, oxidaton state II is not particularly important. However, from Cr to Cu, this state is significant and, for the later elements, is dominant for most compounds. Lower oxidation states must usually be stabilised by π-acceptor ligands, the main exception being Cu^I which owes its considerable stability to the ion's electronic configuration, d^{10}. Oxidation state 0 is dominated by metal carbonyl chemistry, and many organometallic species derived from metal carbonyls have oxidation states of 0, I or II; some may even have negative formal oxidation states (–I, –II, –III).

5.1 Metal carbonyls

Transition metal carbonyls are very useful and relatively easily available starting materials for the synthesis of many other types of low-valent metal complexes.

Some of the commonly encountered first-row binary transition metal carbonyls are shown in Fig. 5.1. Most are prepared by reduction of metal halides, acetates or β-diketonates under pressure of CO gas (*reductive carbonylation*). Exceptions are $[Fe(CO)_5]$ and $[Ni(CO)_4]$ which can be generated under very mild conditions, and $[Fe_2(CO)_9]$ which is formed by UV irradiation of solutions of $[Fe(CO)_5]$ in glacial acetic acid. Trinuclear $[Fe_3(CO)_{12}]$ is obtained by oxidative acidification of an iron carbonylate ion (see later), and $[Co_4(CO)_{12}]$ can be produced by gentle thermolysis of $[Co_2(CO)_8]$. Cluster compounds with higher numbers of metal atoms are also known.

All of these compounds are diamagnetic with the exception of $[V(CO)_6]$. The structures of the dinuclear and higher nuclearity species have two unusual features: bridging carbonyl groups and metal–metal bonds.

All of them, again with the exception of $[V(CO)_6]$, obey the *18-electron rule*. There are a few simple rules which must be observed when using 18-electron counting procedures:

(a) all of the of the atoms and ligands should be treated as electrically neutral; this avoids the complication of assigning formal oxidation states to the metal and deciding on whether the ligand is anionic, neutral or cationic;

(b) in a metal–metal bond, electrons should be equally partitioned, *i.e.*, for M–M each metal atom is allocated one e^-, for M=M, two e^-, and for M≡M, three e^-;

> The *18-electron rule* states that thermodynamically stable transition metal carbonyls and organometallics are formed when the sum of the metal valence electrons ($3d$ and $4s$) plus the electrons supplied by the ligands, adjusted for overall charge, totals 18.

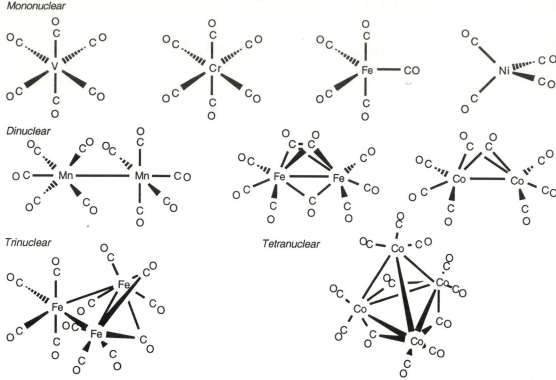

Fig. 5.1 Some commonly encountered binary first-row metal carbonyls

Table 5.1 Ligands and the 18-electron rule (n = no. of electrons donated)

n	Ligand
1	halide, hydride, alkyl, σ-aryl
2	CO, PR_3, $P(OR)_3$, monoalkene
3	NO, cyclopropenyl, η^3-allyl, any other enyl
4	η^4-cyclobutadiene (C_4H_4)
5	η^5-cyclopentadienyl, any other dienyl
6	η^6-arene, tri-alkene
7	η^7-tropylium $([C_7H_7]^+)$
8	η^8-cyclooctatetraene or η^{10}-$[C_8H_8]^{2-}$

(c) bridging ligands are apportioned electrons according to the type of bridge: for CO bridging two metals, each metal receives one e^-; for halide or SR bridging two metals, one metal receives one e^- and the other two e^- {M–X→M};

(d) each C atom in a delocalised hydrocarbon ligand which is directly interacting with a metal donates one e^- per atom;

(e) overall charge on the complex is accounted for by adding one e^- to the total valence electron (VE) count for each negative charge, and deducting one e^- for each positive charge.

The number of electrons allocated to particular ligands for the purposes of electron counting is shown in Table 5.1.

The operation of the rule can be checked against the compounds shown in Fig. 5.1 (Table 5.2). The only exception to the rule is $[V(CO)_6]$, which is paramagnetic (one unpaired electron) and presumably cannot attain the expected electronic configuration because this is prevented sterically. To avoid the complication of assigning oxidation state, it is most convenient to treat the metal as having no formal charge (in its zero-oxidation state), the ligands as electrically neutral (see Table 5.1) and to account for the overall charge on the complex by assuming each negative charge adds one unit and each positive charge deducts one unit from the total electron count.

Table 5.2 Application of the 18-electron rule to binary metal carbonyls[a]

[Cr(CO)$_6$] 6CO = 12e^-; Cr^0 = 6e^- **Total = 18e^-**	[Fe(CO)$_5$] 5CO = 10e^-; Fe^0 = 8e^- **Total = 18e^-**	[Ni(CO)$_4$] 4CO = 8e^-; Ni^0 = 10e^- **Total = 18e^-**
[Mn$_2$(CO)$_{10}$] 5CO = 10e^-; Mn^0 = 7e^- Mn–Mn = 1e^- **Total = 18e^-**	[Fe$_2$(CO)$_9$] 3CO = 6e^-; 3 μ-CO = 3e^- Fe^0 = 8e^-; Fe–Fe = 1e^- **Total = 18e^-**	[Co$_2$(CO)$_8$] 4CO = 8e^-; Co^0 = 9e^- Co–Co = 1e^- **Total = 18e^-**
[Fe$_3$(CO)$_{12}$] *Fe(CO)$_4$:* 4CO = 8e^- Fe^0 = 8e^-; 2Fe–Fe = 2e^- **Sub-total 18e^-** *Fe$_2$(CO)$_8$:* 3CO = 6e^- 2 μ-CO = 2e^-; Fe^0 = 8e^- 2Fe–Fe = 2e^- **Sub-total 18e^-**	[Co$_4$(CO)$_{12}$] *Apical Co(CO)$_3$:* 3CO = 6e^-; Co^0 = 9e^- 3Co–Co = 3e^- **Sub-total 18e^-** *Basal Co$_3$(CO)$_9$:* 2CO = 4e^- 2 μ-CO = 2e^-; Co^0 = 9e^- 3Co–Co = 3e^- **Sub-total 18e^-**	[V(CO)$_6$] 6CO = 12e^-; V^0 = 5e^- **Total 17e^-**

[a] For oligonuclear species, electron-counting is based on single metal centres.

Although low-valent compounds with ligands other than CO have not yet been discussed, it is appropriate to list in Table 5.1 the electrons contributed by such ligands, in the neutral, positively or negatively charged state, relevant to the application of the 18-electron rule.

There are, however, some important exceptions to the 18-electron rule which are important in metal carbonyl and organometallic chemistry. The first is the occurrence of stable compounds which have a formal 16-valence electron configuration, *e.g.*, [Ni(CN)$_4$]$^{2-}$ and [Ni(PPh$_3$)$_3$]. Sometimes reactive species with 14-electron configurations can also be detected. This phenomenon is of considerable relevance to mechanisms of homogeneous catalysis. The second has already been referred to, *viz.* the occasional observation of compounds having 17- or even 19-electron configurations. Examples of stable 17-electron compounds include [V(CO)$_6$] and [Mn(CO)$_5$(PEt$_3$)], and of 19-electron species, [Co(η^5-C$_5$H$_5$)$_2$].

While the majority of metal carbonyl and organometallic compounds obey the 18-electron rule, many coordination compounds containing ligands such as NH$_3$, water, oxalate, *etc.*, do not. The reason for this in octahedral complexes can be seen by studying the molecular orbital diagram Fig. 1.11 (p.16), which shows that the crystal field splitting, Δ_{oct}, is larger when there is π bonding between the metal and the ligand, such as occurs in the carbonyl and related species, than when only σ bonding is involved, as in NH$_3$ and H$_2$O complexes. This is entirely consistent with the spectrochemical series where ligands like CO, PR$_3$, alkenes, CN$^-$, alkyl and aryl all exert a stronger crystal field than H$_2$O, NH$_3$, and halide.

For octahedral compounds in which Δ_{oct} is relatively small and the ligands are at the lower end of the spectrochemical series, t_{2g} is non-bonding and can be occupied by 0–6 electrons while the e_g^* level is weakly antibonding and can be occupied by 0–4 electrons. Hence between 12 and 22 valence electrons can be accommodated and so the 18-electron rule is not

For an explanation of the η nomenclature, see Section 5.6, p.70.

Isonitriles, RNC, are isoelectronic with CO and form complexes closely similar to metal carbonyls: [Cr(NCR)$_6$], [Fe(CNR)$_5$], [Fe$_2$(CNR)$_9$], [Co$_2$(CNR)$_8$] and [Ni(CNR)$_4$]. These compounds may be prepared by reductive methods, or by displacement of CO from metal carbonyls. Isonitriles can also stabilise metals in oxidations states higher than 0, for example, [Mn(CNR)$_6$]$^{2+}$, for which there is no binary carbonyl counterpart. Isonitriles are stronger σ donors than CO: they are also weaker π acceptors but this is less important in oxidation states of II and higher. Like CO, isonitriles can act as bridging ligands, as in [(RNC)$_3$Fe(μ-CNR)$_3$Fe(CNR)$_3$], and are involved in group migratory reactions with metal-alkyls, giving the moiety {M–C(=NR)R}.

Table 5.3 The 18-electron rule and transition metal complexes

Typical transition metal complex	M + nL + charge
$[VCl_6]^{2-}$	5+6+2=13
$[Cr(acac)_3]$	6+9+0=15
$[FeF_6]^{3-}$	8+6+3=17
$Co(NH_3)_6]^{3+}$	9+12−3=18
Complexes with π-acceptor ligands	
$[V(CO)_6]^-$	5+12+1=18
$[Mn(CO)_5Br]$	7+10+1=18
$[Co(CO)_4Me]$	9+8+1=18
$[Ni(PMe_3)_4]$	10+8=18

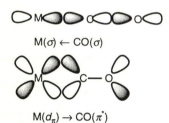

$M(\sigma) \leftarrow CO(\sigma)$

$M(d_\pi) \rightarrow CO(\pi^*)$

Fig. 5.2 Donation from CO to a metal *d* or *p* acceptor orbital and back-donation from a metal *d* orbital to CO antibonding π^* orbitals

Evidence that the lone pair on C is weakly antibonding can be obtained from the increase inCO stretching frequency from 2143 cm^{-1} in free CO to 2164 cm^{-1} in the adduct $H_3B{\leftarrow}CO$. The increase in C–O bond order in the adduct reflects the decrease in the antibonding ionfluence of the lone pair.

obeyed. Examples are shown in Table 5.3. Tetrahedral complexes, because of their small crystal field splitting Δ_{tet}, also belong to this group.

However, when Δ_{oct} is large, as in complexes with ligands at the higher end of the spectrochemical series such as CO, PR$_3$, alkenes, *etc.*, the t_{2g} becomes bonding due to the interaction with π and π^* orbitals of the ligands, and should be fully occupied by six electrons. The $e_g{}^*$ level is strongly antibonding and should remain empty. Hence the 18-electron rule is obeyed, unless there is a particular steric reason which prevents attainment of the 18-electron valence shell, as in $[V(CO)_6]$.

The thermodynamic stability of metal carbonyl derivatives owes very little to the ability of CO to act as a Lewis base. It owes a great deal, however, to the ability of CO to act as a π-acceptor ligand. This is shown in Fig. 5.2. The lone electron pair on the C atom is weakly σ antibonding with respeco to CO and so forms a weak M–C bond in carbonyl complexes. However, the two π^* orbitals of CO have the correct symmetry and energy to interact with appropriate $d(\pi)$ orbitals of the transition metal. The effect of this type of bonding on the behaviour spectroscopically and structurally of the CO group is very important. First, removal of σ^*-antibonding electron density from the CO group as the first 'act' of σ bonding to the metal has little effect, perhaps causing a marginal increase in the C–O bond length. This is difficult to monitor since there are very few stable metal carbonyl complexes in which CO functions solely as a σ-donor ligand. However, back-donation from the metal to the carbonyl ligand leads to population of the π^* (antibonding) orbitals, and to an effective repulsion between the C and the O atoms. In other words, the C–O bond increases in comparison to its separation in the simple molecule.

Evidence to support the concept of back-donation from metal to CO is obtained partly from crystallographic structure determinations. The distance between the C and O atoms in moving from bond order 3 to 2 is barely detectable by X-ray diffraction methods; the C–O distance in free carbon monoxide is 1.13 Å whereas in metal carbonyls it averages 1.15 Å. However, inspection of the M–C distances is more revealing since there is a progression from M–C to M=C which is accompanied by a bond shortening from 0.3–0.4 Å.

The most significant evidence for back-donation comes from infrared spectroscopy. Carbon monoxide is dipolar and therefore gives rise to an easily detectable stretching vibration (at 2143 cm^{-1} in the free molecule). To a first approximation it is reasonable to regard C–O vibrations as being independent of other vibrations in the molecule, an assumption which cannot be made for M–C vibrations. In the IR spectra of metal carbonyls, v_{CO} usually falls at or below 2100 cm^{-1}, indicating a significant decrease in the C–O bond order consistent with back-donation.

However, the exact position of v_{CO} is strongly dependent on the bonding mode of carbon monoxide in the molecule, on the charge on the compound and on the electronic properties of co-ligands in the molecule. The effect of bonding mode is shown in Fig.5.3, and of charge in Table 5.4.

Table 5.4 Effect of charge on CO stretching frequency in isoelectronic series

$[M(CO)_6]^z$	ν_{CO} (cm^{-1})	$[M(CO)_4]^z$	ν_{CO} (cm^{-1})
$[V(CO)_6]^-$	1860	$[Fe(CO)_4]^{2-}$	1790
$[Cr(CO)_6]$	2000	$[Co(CO)_4]^-$	1890
$[Mn(CO)_6]^+$	2090	$[Ni(CO)_4]$	2060

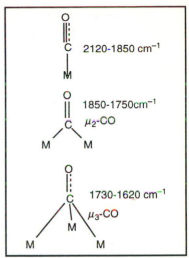

Fig 5.3. Carbonyl bonding modes

Increasing negative charge on the complex leads to an expansion of the metal d orbitals with a concomitant increase in back-donation to the π^* orbitals of the CO groups. Hence ν_{CO} decreases. Conversely, an increase in positive charge leads to a contraction of the d orbitals and a reduction of back-donation and an increase in ν_{CO}.

The effect of co-ligands L in substituted metal carbonyls $[M(CO)_nL_m]$ on CO stretching frequencies reflects the σ donor and π acceptor properties of L. Ligands *trans* to CO will interact with the same metal d orbital. If L is a poor π acceptor but a strong σ donor, electron density will be increased on the metal which will be dispersed into the π^* orbitals of CO. The effect will be a decrease in C–O bond order and a reduction in ν_{CO}. However if L is a good π acceptor, then there will be less need for back-donation to CO. From CO stretching frequency data in a series of metal carbonyls, an order of π-acceptor strength for many ligand types can be constructed: NO > CO > RNC > PF_3 > PCl_3 > $P(OR)_3$ > $P(aryl)_3$ > $P(alkyl)_3$ > RCN > NH_3, py.

Reactions of metal carbonyls

These may be categorised as substitution, reduction to form carbonylate ions, oxidation, and attack on coordinated CO.

Substitution

Carbonyl groups can be substituted by other ligands (tertiary phoshines, alkenes, *etc.*) thermally or photochemically in a dissociative process (see Section 3.3). Substitution can proceed to the point that there are equal numbers of ligands L and carbonyl groups, but replacement of the remaining CO groups is virtually impossible because of increasingly strong M–CO back-donation. Monosubstitution is often achieved photochemically, and if the M–L bond in $[M(CO)_nL]$ is labile, displacement by other groups is easily effected.

$$2[Fe(CO)_5] + 3PPh_3 \longrightarrow [Fe(CO)_4(PPh_3)] + [Fe(CO)_3(PPh_3)_2]$$

$$[Cr(CO)_6] + \textit{cis}\text{-cyclooctene} \longrightarrow \bigcirc\!\!\!-\!-\!-\cdot Cr(CO)_5 \xrightarrow{\text{L}} [Cr(CO)_5L]$$

Photolytic cleavage of the M–M bond in $[Mn_2(CO)_{10}]$ and $[Co_2(CO)_8]$ is a useful pathway to asymmetrically substituted and even paramagnetic species.

$$[Co_2(CO)_8] \xrightarrow{h\nu} 2\ [Co(CO)_4] \xrightarrow{L} [Co(CO)_4] + [Co(CO)_3L] \longrightarrow [Co_2(CO)_7L]$$

$$[Mn_2(CO)_{10}] \xrightarrow{h\nu} 2\ [Mn(CO)_5] \xrightarrow{PPh_3} [Mn(CO)_4(PPh_3)]$$

In contrast to the other mononuclear carbonyl complexes, $[V(CO)_6]$ undergoes very rapid substitution by an associative process, affording paramagnetic $[V(CO)_5L]$.

Reduction to form carbonylate ions

A very large number of anionic metal carbonyl species, the carbonylate ions, are known. They may be generated in a number of ways, the most important being nucleophilic attack on a coordinated CO group (*e.g.*, Scheme 5.1), followed by loss of CO_2, or by reaction with an alkali metal in an ether or an alkali metal amalgam.

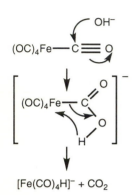

$$[Fe(CO)_5] + OH^- \longrightarrow [Fe(CO)_4H]^- + CO_3^{2-} \xrightarrow{-H^+} [Fe(CO)_4]^{2-}$$

$$[Fe_3(CO)_{12}] + OH^- \longrightarrow [Fe_3(CO)_{11}]^{2-} + [Fe_3(CO)_{11}H]^- + CO_3^{2-}$$

$$[Mn_2(CO)_{10}] + 2\ Li \longrightarrow 2\ Li[Mn(CO)_5]$$

$$[Co_2(CO)_8] + 2Na/Hg \xrightarrow{THF} 2\ Na[Co(CO)_4]$$

Scheme 5.1 Nucleophilic attack by OH^- on $[Fe(CO)_5]$

The formal oxidation states in these carbonylate ions may be $-I$ as in $[Co(CO)_4]^-$, or $-II$ as in $[Fe(CO)_4]^{2-}$. Reduction of the carbonyls by much stronger reagents such as sodium in liquid ammonia, lithium naphthalenide or sodium hexamethylphosphoramide affords the highly reactive 'super-reduced' carbonyl metallates. Examples include $Na_3[Mn(CO)_4]$ (formally containing Mn^{-III}) and $Na_4[Cr(CO)_4]$ (formally containing Cr^{-IV}). Their CO stretching frequencies occur in the region $1800–1450\ cm^{-1}$.

Carbonylate ions are basic, and can be readily protonated to form *metal carbonyl hydrides*, $[M(CO)_nH_m]$. The hydrides may also be formed by direct reaction with hydrogen:

$$[Co_2(CO)_8] + H_2 \rightarrow 2\ [Co(CO)_4H]$$

or by reaction of carbonyl halides (see below) with BH_4^-:

$$[Mn(CO)_5Br] + BH_4^- \rightarrow [Mn(CO)_5H]$$

In water $[Co(CO)_4H]$ is a strong acid ($pK_a = 1$), comparable to H_2SO_4, whereas $[Fe(CO)_4H_2]$ ($pK_a = 4.7$) and $[Mn(CO)_5H]$ ($pK_a = 7$) are weaker acids. The hydrides are extremely reactive, undergoing 'insertion' reactions with alkenes and alkynes forming metal-alkyl and related species:

$$[Mn(CO)_5H] + C_2F_4 \rightarrow [Mn(CO)_5(CF_2CF_2H)]$$

The carbonylate ions also react with alkyl and acyl halides forming *metal carbonyl alkyl and acyl* derivatives, the carbonylate ion acting as a powerful nucleophile towards the organic halide:

$$[Mn(CO)_5]^- + MeI \rightarrow [Mn(CO)_5Me] + I^-$$

$$[Mn(CO)_5]^- + EtCOCl \rightarrow [Mn(CO)_5\{C(O)Et\}] + Cl^-$$

Acyl derivatives may also be formed by reaction of a metal alkyl with carbon monoxide in a process which may occur spontaneously or under pressure of CO. The interconversion between metal-alkyl and metal-acyl bonding is of great importance in the exploitation of carbonyl derivatives in synthesis, mechanistic information being obtained from an IR study of the reactions of $[Mn(CO)_5Me]$ with ^{13}CO (Scheme 5.2).

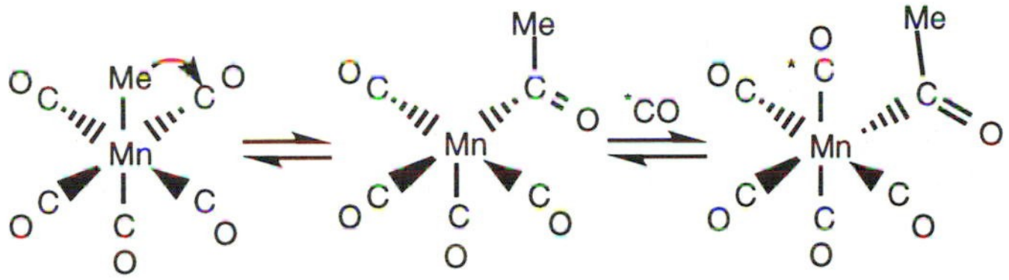

Scheme 5.2 CO insertion into a Mn–methyl bond (group migration)

The reaction is popularly known as an '*insertion*' into the metal–carbon bond, but in fact the methyl group migrates to an adjacent carbonyl position, and the overall process is more correctly termed a 'group migration' reaction.

Alkyls and acyls derived from $[Fe(CO)_4]^{2-}$ have significant applications in organic synthesis and $Na_2[Fe(CO)_4]$ is sometimes known as *Collman's reagent*. Important alkyl and acyl iron intermediates, obtained by reaction of $[Fe(CO)_4]^{2-}$ with RX or RC(O)X, are shown in Fig. 5.4. These react with O_2, alkyl halides and halogens affording carboxylic acids, ketones and acyl halides, respectively.

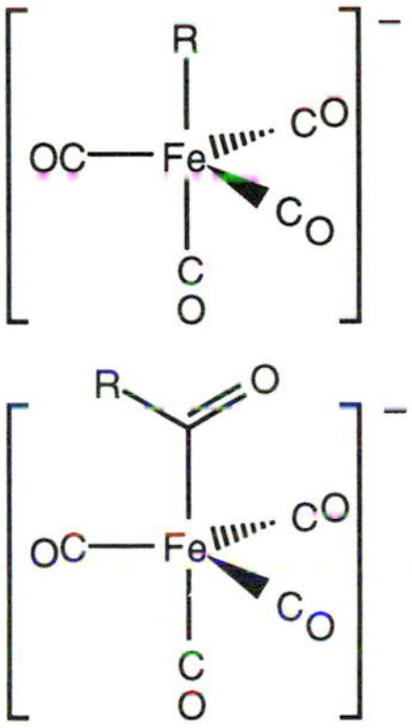

The particular ability of $[Co_2(CO)_8]$ to activate molecular hydrogen, forming $[Co(CO)_4H]$, and of this hydride to undergo 'insertion' reactions with alkenes forming intermediary cobalt alkyls and acyls, is exploited in industrially important *hydroformylation* processes. The process operates at elevated temperatures under high pressures of CO/H_2 mixtures and is often referred to as the '*oxo process*'. The hydroformylation process formally represents addition of H and HCO across a double bond (RCH=CH$_2$ + CO + $H_2 \rightarrow$ RCH$_2$CH$_2$CHO), and acyls $[Co(CO)_4\{C(O)R\}]$ are important intermediates.

Fig. 5.4 Alkyl and acyl derivatives of $[Fe(CO)_4]^{2-}$ important in organic syntheses

Oxidation of metal carbonyls

The most common oxidising agents are halogens. Under relatively mild conditions, one or two CO groups may be displaced by a halogen, affording metal carbonyl halides, but under more vigorous conditions, complete oxidative decarbonylation occurs. This latter process is sometimes exploited for the preparation of reactive anhydrous metal halides.

Metal carbonyl halides, an important and useful class of starting material in the preparation of many derivatives, including organometallics, can be prepared by direct reaction of the carbonyl or by metal–metal bond cleavage by halogen, by substitution of CO by halide ion and, occasionally, by reaction of CO with metal halides. Examples of such reactions are shown below:

$[Fe(CO)_5] \xrightarrow{\text{Br}_2} \textit{cis-}[Fe(CO)_4Br_2] + CO$

$[Mn_2(CO)_{10}] \xrightarrow{\text{I}_2} 2\,[Mn(CO)_5I] \xrightarrow{\text{heat}}$

$[Cr(CO)_6] \xrightarrow{\text{Cl}^-} [Cr(CO)_5Cl]^- + CO$

$2\,[CuCl_2]^- \xrightarrow{\text{CO}} [Cu(CO)(\mu\text{-Cl})]_2 + 2\,Cl^-$

The halide can be displaced by other nucleophiles, such as CN^-, $[C_5H_5]^-$ or H^- giving cyano-, η-cyclopentadienyl or hydrido metal carbonyl derivatives. Halide may also be abstracted under CO pressure to form carbonyl cations, *e.g.*, by reacting $[Mn(CO)_5Cl]$ with $AlCl_3$, which gives $[Mn(CO)_6]^+[AlCl_4]^-$, isoelectronic with $[Cr(CO)_6]$.

Attack on coordinated CO

The hydrido-carbonylate ion $[Fe(CO)_4H]^-$ is obtained by nucleophilic attack of OH^- on a coordinated CO in $[Fe(CO)_5]$ (Scheme 5.1, p.62). If methoxide is used in place of OH^-, intermediates such as $[Fe(CO)_4\{C(O)OMe\}]^-$ can be isolated. A similar reaction occurs with $[Fe_3(CO)_{12}]$ when $[Fe_3(CO)_{11}\{C(O)OMe\}]^-$ is formed.

Nucleophilic attack by carbanions on coordinated CO affords anionic metal acylate derivatives. A typical reaction involving $[Fe(CO)_5]$ was illustrated in Scheme 5.1, but the reactions of $[Cr(CO)_6]$ are also important:

$$[Cr(CO)_6] + LiR \rightarrow [Cr(CO)_5\{C(O)R\}]^-$$

Methylation of this derivative with $[Me_3O][BF_4]$ affords the *carbene* complex $[Cr(CO)_5\{C(OMe)R\}]$. The carbene ligand is also susceptible to nucleophilic attack, particularly by amines NHR_2, the OMe group being replaced by NR_2. Reaction of these carbene complexes with BX_3 (X = Cl, Br, I), in a reaction designed to displace the alkoxy group by X, led instead to *trans*-$[Cr(CO)_4(CR)X]$, a metal *carbyne* complex (Fig. 5.5).

The carbene ligand is a weaker π acceptor and a stronger σ donor than CO and may be regarded as a two-electron donor, *i.e.*, like a conventional Lewis base. The C atom is sp^2-hybridised, and the Cr–C(carbene) bond is significantly shorter than an M–C single bond, but longer than the M–C(carbonyl) bond. This shows that there is significant multiplicity in the M–C(carbene) interaction (Fig. 5.6). In terms of valence-bond structures, the metal–carbene interaction may be depicted as $\{M=C(OR)R\}$.

The Cr–C(carbyne) bond length is shorter than those in carbene complexes, and the Cr–C–R bond is linear. There is obviously overlap between a metal *d* orbital and the two *p* orbitals on the unique C atom (Fig. 5.7). The carbyne ligand is regarded as a three-electron donor, the interaction being depicted in valence-bond terms as $\{M\equiv CR\}$.

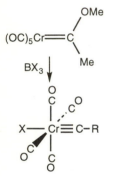

Fig. 5.5 Conversion of a chromium carbene to a carbyne

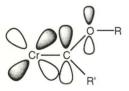

Fig. 5.6 Bonding in low-oxidation state carbene

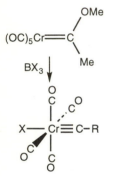

Fig. 5.7 Bonding in carbyne complexes showing interaction of one d_π orbital with one of the *p* orbitals on the carbyne C atom

5.2 Nitrosyl complexes

As a molecule, NO has one more electron than CO, in a π^* orbital. The nitrosyl group may be regarded as either a three- or a one-electron donor. In the former mode, the π^* electron is notionally donated to the metal, thereby reducing it, and then σ donation occurs from the N atom lone pair to the metal. The M–N–O bond angle in this arrangement is essentially linear, the nitrosyl group being formally regarded as coordinated NO^+ which is isoelectronic with CO. Thus the NO group formally contributes three electrons to the metal. In the other bonding mode, the metal notionally donates an electron to the NO thereby itself becoming oxidised, the nitrosyl group becomes formally NO^-. However, the N atom rehybridises to sp^2 and a lone pair of electrons is accommodated on the N atom of the nitrosyl group and does not interact with the metal. NO^- then behaves formally like chloride or alkyl ligands, but contributes overall only one electron to the metal–NO interaction. The occurrence of one or other of these bonding modes is dependent on the electronic structure of the particular compound (see below).

There is only one binary nitrosyl complex, $[Cr(NO)_4]$, but a number of mixed carbonyl nitrosyl complexes are known. These include a group which is isoelectronic and isostructural with $[Ni(CO)_4]$, *viz.*, $[Co(CO)_3(NO)]$, $[Fe(CO)_2(NO)_2]$, $[Mn(CO)(NO)_3]$ and $[Cr(NO)_4]$, and the anion $[Fe(CO)_3(NO)]^-$. $[Mn(CO)_4(NO)]$ is formally isoelectronic with $[Fe(CO)_5]$. There are many other types of nitrosyl complexes, including a series of nitrosyl cyanides, *e.g.*, $[Cr(CN)_5(NO)]^{3-}$ and $[Fe(CN)_5(NO)]^{2-}$ (the nitroprusside ion), nitrosyl halides, *e.g.*, $[Fe(NO)_2(\mu\text{-}X)]_2$ and $[Co(NO)_2(\mu\text{-}X)]_2$, and organometallic species, *e.g.*, $[Ni(\eta\text{-}C_5H_5)(NO)]$, $[Mn(\eta\text{-}C_5H_5(CO)(NO)(PPh_3)]^+$ and $[\{Cr(\eta\text{-}C_5H_5)(NO)(\mu\text{-}NO)\}_2]$.

Nitrosyl complexes can be prepared by direct reaction of NO or NO^+ with metal carbonyls or other low-oxidation state compounds. They may also be prepared by reaction of appropriate metal salts with hydroxylamine in basic conditions or by reduction *in situ*, often by CO, of nitrite ion.

$$[Mn_2(CO)_{10}] + 2NO \rightarrow 2[Mn(CO)_4(NO)] + 2CO$$

$$[Mn(CN)_6]^{3-} + NH_2OH + OH^- \rightarrow [Mn(CN)_5(NO)]^{3-} + CN^- + NH_3 + 2H_2O$$

$$[Fe(CO)_5] + NO_2^- \rightarrow [Fe(CO)_3(NO)]^- + CO + CO_2$$

Like metal carbonyl chemistry, the IR stretching frequency of metal-coordinated NO is extremely sensitive to bonding mode. Free NO and the nitrosonium ion have ν_{NO} at 1876 and 2205 cm^{-1}, the increase of ν_{NO} in NO^+ being caused by loss of the antibonding π^* electron. There is an increase in the N–O bond order from 2.5 to 3. In metal nitrosyls, ν_{NO} falls in the range 1900–1600 cm^{-1} for terminal linear M–N–O groups, but like CO, NO can also act as a bridging ligand, with appropriate reductions in ν_{NO} (Fig. 5.8).

Complexes containing 'bent' nitrosyl groups are not very common and their prediction is difficult. Generally, however, when a complex is relatively electron rich, and the metal electronic configuration is likely to exceed 18,

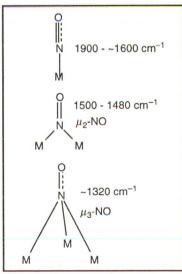

Fig. 5.8 NO stretching frequency range

then the M–N–O group can distort, v_{NO} occuring in the range 1600–1500 cm^{-1} which is in the range of bridging NO.

As mentioned earlier, transition metal nitrosyls are substituted in an associatively activated process (Section 3.3, Scheme 3.8). This is in contrast to the substitution chemistry of metal carbonyls which is mainly dissociatively activated. Implicit in nitrosyl compound substitution is the 'bent' M–N–O group in which an electron pair is formally withdrawn from the metal and located on a rehybridised N atom on NO, now formally NO$^-$, a one-electron donor. In this way, an orbital is released at the metal to accept the incoming nucleophile. A model of this behaviour is illustrated in Scheme 5.3.

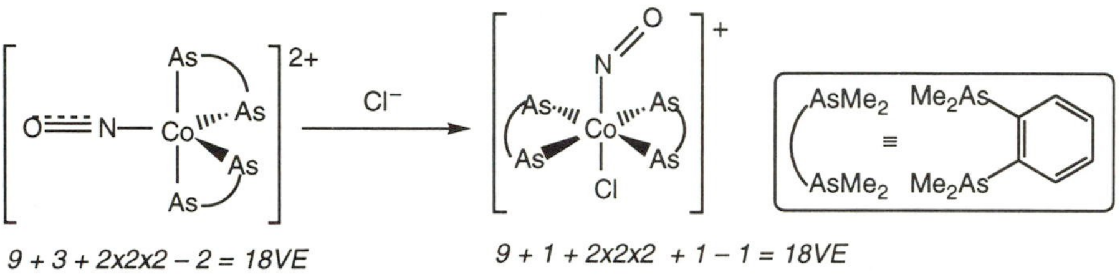

$$9 + 3 + 2\times2\times2 - 2 = 18VE \qquad\qquad 9 + 1 + 2\times2\times2 + 1 - 1 = 18VE$$

Scheme 5.3 Relationship between linear and 'bent' M–N–O

Linearly coordinated NO, like CO, is susceptible to nucleophilic attack in certain cases. Hydroxide ion addition to $[Fe(CN)_5(NO)]^{2-}$ affords the nitro-species $[Fe(CN)_5(NO_2)]^{4-}$, and other nucleophiles, *e.g.*, S^{2-}, SO_3^{2-}, and certain carbanions, also attack the NO group, forming nitroso ligands with characteristic colours. 'Bent' NO, because of the availability of the lone pair, undergoes electrophilic addition:

$$[Co(NO)(diars)_2Cl]^+ + H^+ \rightarrow [Co(HNO)(diars)_2Cl]^{2+}$$

(diars = o-(Me$_2$As)$_2$C$_6$H$_4$). NO 'insertion' reactions are also known, but are not as important as their CO analogues.

5.3 Tertiary phosphine complexes

Ligands of the type PX$_3$, where X = halide, alkyl, Ph, OR, *etc.*, are capable of π and/or σ bonding. In this they are different to CO, which has little σ-bonding capability, but somewhat like CNR. In low oxidation state compounds, of course, tertiary phosphines use both σ and π bonding, the latter involving empty $d\pi$ orbitals on the P atom (Fig. 5.9).

The Lewis basicity of PR$_3$ ligands varies according to R. Some measure of this can be obtained from IR studies of particular CO stretching vibrations in compounds such as $[Ni(CO)_3(PR_3)]$. From this study an order of increasing π-acceptor ability for PR$_3$ can be obtained: PBu$^t_3$ < PMe$_3$ < P(OMe)$_3$ < P(OAr)$_3$ < PCl$_3$ < PF$_3$. In general, trialkyl phosphines are stronger donors but poorer π acceptors than triaryl phosphines, but this order is reversed in P(OR)$_3$ and P(OAr)$_3$. Also, PCl$_3$ is a better σ donor and poorer π acceptor than PF$_3$. This behaviour reflects the inductive effect and the electronegativity of the R groups which, in turn, influence the availability of the P d orbitals for back-donation from the metal. In other words, there is a

Fig. 5.9 π donation from M to PR$_3$

fine synergistic balance between σ donor and π acceptor behaviour in the M–PR$_3$ bond. Supporting evidence for the general order and the particular behaviour of tertiary phosphine substituents is obtained from the observation that the Cr–P bond in [Cr(CO)$_3$(PPh$_3$)] is 0.11 Å longer than in [Cr(CO)$_5${P(OPh)$_3$}]: PPh$_3$ is a stronger σ donor, although poorer π acceptor, than P(OPh)$_3$, but there is greater back-donation from the Cr to the latter than the former.

Trialkyl-, as well as triaryl-phosphines, can stabilise Ni^0, as in [Ni(PMe$_3$)$_4$]. This facility is not available to NH$_3$ or nitriles, RCN, because they do not have π-acceptor orbitals at energies appropriate to interact with metal π-donor orbitals. PF$_3$ is comparable to CO as a ligand, and can support many compounds analogous to binary carbonyls, *e.g.*, [Cr(PF$_3$)$_6$], [Fe(PF$_3$)$_5$] and [Ni(PF$_3$)$_4$].

At least as important as the electronic properties of particular PR$_3$ ligands are steric factors. These may even dominate the stereochemistries and structures of particular complexes, and in so doing may exert a powerful influence on the stability or lability of particular complexes, particularly in catalytic processes. The steric requirement of a particular PR$_3$ ligand is usually expressed by *Tolman's cone angle*, θ, as defined in Fig. 5.10. The cone is one that can just enclose the van der Waals surface of all the ligand atoms over all rotational orientations about the M–P bond. An increase in cone angle in PR$_3$ tends to favour lower coordination numbers in complexes, the formation of less sterically crowded isomers, and increased rates and equilibria in dissociative reactions. An example of this last effect is demonstrated in the equilibrium:

$$[Ni(PR_3)_4] \rightleftharpoons [Ni(PR_3)_3] + PR_3$$

The equilibrium constant K varies from >1 for L = PPh$_3$ to <10^{-10} for L = P(OEt)$_3$.

Tertiary phosphine complexes of metals in oxidation state II are usually prepared by reaction of divalent metal salts with PR$_3$.

$$CoBr_2 + 3PMe_3 \rightarrow [Co(PMe_3)_3Br_2]$$

$$FeCl_2 + 2Me_2PCH_2CH_2PMe_2 \rightarrow [Fe(Me_2PCH_2CH_2PMe_2)_2Cl_2]$$

Many of these complexes do not obey the 18-electron rule, *e.g.*, [Ni(PEt$_3$)$_2$(NCS)$_2$] (diamagnetic, planar, 16VE) and [Ni(PPh$_3$)$_2$Br$_2$] (paramagnetic, tetrahedral, 16VE). Complexes in lower oxidation states are prepared either by formation of a phosphine complex with a metal in a medium oxidation state, followed by reduction in the presence of an excess of PR$_3$, or by displacement of labile hydrocarbon ligands by PR$_3$.

$$[Fe(PF_3)_4Cl_2] + PF_3 + 2Na/Hg \rightarrow [Fe(PF_3)_5] + 2NaCl$$

$$[Ni(\eta\text{-}C_5H_5)_2] + 4\,P(OPh)_3 \rightarrow [Ni\{P(OPh)_3\}_4]$$

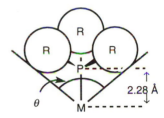

Fig. 5.10 Tolman's cone angle, arbitrarily setting M–P at 2.28 Å

Cone angles for typically encountered PR$_3$ ligands are PF$_3$ 104°, P(OMe)$_3$ 107°, P(OEt)$_3$ 109°, PMe$_3$ 118°, P(OPh)$_3$ 128°, PPh$_3$ 145°, PBu$^t_3$ 182°.

An important application of nickel tertiary phosphine complexes is the hydrocyanation of alkenes. Butadiene or acrylonitrile is converted to adiponitrile or succinonitrile which is reduced to hexamethylenediamine or 1,4-diaminobutane, important intermediates in the synthesis of various types of nylon.

5.4 Alkene complexes

Alkene complexes are obtained by substitution of CO in metal carbonyls or of other labile ligands in low-oxidation state compounds, by reductive methods, or by reaction of metal atoms with particular alkenes (Scheme 5.4).

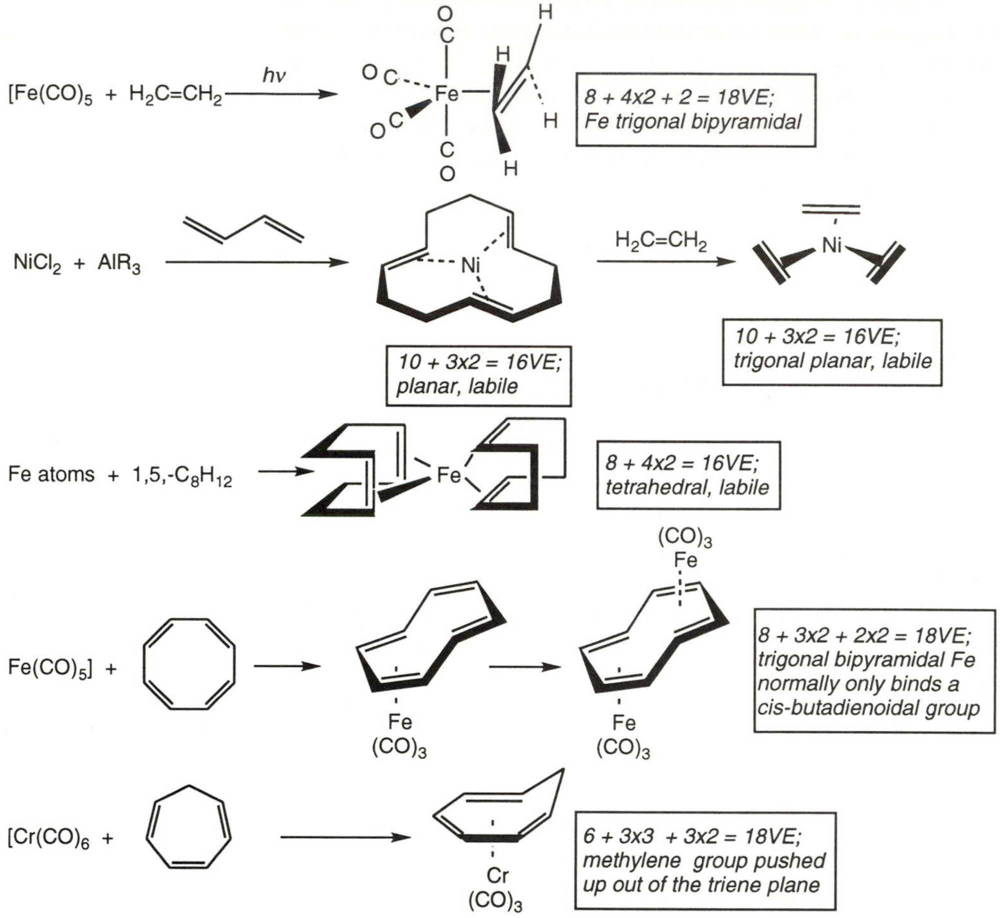

$[Fe(CO)_5 + H_2C=CH_2] \xrightarrow{hv}$

8 + 4x2 + 2 = 18VE; Fe trigonal bipyramidal

$NiCl_2 + AlR_3 \longrightarrow$ $\xrightarrow{H_2C=CH_2}$

10 + 3x2 = 16VE; planar, labile

10 + 3x2 = 16VE; trigonal planar, labile

Fe atoms + 1,5,-C_8H_{12} $\longrightarrow$

8 + 4x2 = 16VE; tetrahedral, labile

$Fe(CO)_5] +$ $\longrightarrow$

8 + 3x2 + 2x2 = 18VE; trigonal bipyramidal Fe normally only binds a cis-butadienoidal group

$[Cr(CO)_6 +$ $\longrightarrow$

6 + 3x3 + 3x2 = 18VE; methylene group pushed up out of the triene plane

Scheme 5.4 Synthesis of metal alkene complexes

The bonding between an alkene and a metal is shown in Fig. 5.11. The σ interaction is relatively weak, involving donation of electron density from a π-bonding orbital to an acceptor (d or dp hybrid) orbital on the metal, and this is complemented by back-donation from the metal to a π^* orbital on the alkene.

π Donation from alkene to M.   Back-donation from M to alkene π^* orbital.

Fig. 5.11 Metal–alkene bonding

There are two major consequences of this effective transfer of electron density from a π-bonding to a π^*-antibonding orbital on the alkene. These are (i) a small but usually significant lengthening of the alkene C–C bond which should lead to a decrease in the bond stretching force constant and, therefore, v_{CC}, and (ii) a loss of planarity in the alkene as back-donated charge builds up on the alkene C atoms, so that the metal–alkene interaction increasingly resembles a cyclopropane ring in which one CH_2 group is replaced by the metal fragment (a metallacyclopropane ring).

Dealing first with (i), the C=C bond length in free ethene is 1.35 Å whereas in alkene complexes the C–C distance can fall between this and values close to that of a C–C single bond (1.53 Å; 1.46 Å in $[Fe(CO)_4(C_2H_4)]$). However, while v_{CC} in free ethene is 1623 cm^{-1}, it is 1551 cm^{-1} in $[Fe(CO)_4(C_2H_4)]$ and 1508 cm^{-1} in $[Mn(\eta\text{-}C_5H_5)(CO)_2(C_2H_4)]$. This is consistent with the bonding picture shown above.

Regarding (ii), the C atoms are sp^2-hybridised in the uncoordinated alkene, but on coordination, when significant back-donation occurs, they begin to rehybridise with an increase in p character becoming, in the limit, sp^3. This leads to the distortion shown in Fig. 5.12, an extreme form of back-donation is tantamount to regarding the interaction as $\{L_nM^{2+}\cdots(R_2CCR_2)^{2-}\}$.

In trigonal and planar low oxidation state complexes, the alkene is always located in the metal–ligand plane, almost never orthogonal to it. In trigonal bipyramids, the alkene normally occupies the equatorial plane, and is coplanar with the other equatorial ligands. This arrangement maximises the back-donation between metal and ligand. However, there are d orbitals orthogonal to this ML_2 plane which can interact with the alkene and this facilitates rotation of the alkene about a line from its mid-point to the metal. NMR spectroscopy has confirmed this rotation.

Non-conjugated dienes, trienes, *etc.*, bind to transition metals as though each C=C bond is essentially independent, the interaction being the same as that in a monoalkene described above. When the oligo-olefin can function as a chelating ligand, *e.g.*, as in 1,5-cyclooctadiene, bicyclo-[2.2.4]-heptadiene (norbornadiene) and even cyclooctatetraene, the stability of the complexes is significantly higher than that of comparable monoalkene complexes.

Conjugated dienes and higher olefins usually bind *via* the η^4-*s*-*cis* mode (Fig. 5.13), but in cyclic alkenes such as cycloheptatriene or cyclooctatetraene all the double bonds may not be used in binding. The bonding is generally similar to the other alkenes, but is modified because the cyclic ligands have delocalised π MOs of lower energies and different symmetries compared with their non-conjugated analogues and they are potentially more effective π acceptors.

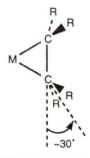

Fig. 5.12 Effect of rehybridisation of alkene C atoms as a result of π back-donation from the metal

Fig. 5.13 *Cis* diene binding to M

Reactions of alkene and oligo-olefin complexes

As in metal carbonyl chemistry, a significant part of alkene complex reactivity is displacement of the olefin by other Lewis bases. This can sometimes be exploited by using labile complexes, *e.g.*, formation of a dinuclear η-allyl nickel complex (see Section 5.6):

$$2[Ni(C_8H_{12})_2] + 2H_2C=CHCH_2Br \rightarrow [\{Ni(\eta^3\text{-}C_3H_5)(\mu\text{-}Br)\}_2]$$

Coordinated alkenes can be attacked by electrophiles or nucleophiles:

$$[Fe(\eta^4\text{-}C_4H_6)(CO)_3] + H^+ + CO \rightarrow [Fe(\eta^3\text{-}C_3H_4Me)(CO)_4]^+$$

$$[Mn(\eta^2\text{-}C_2H_4)(CO)_5]^+ + H^- \rightarrow [Mn(CO)_5(C_2H_5)]$$

Protonation of cyclic diene complexes can afford cationic η^5-dienyl complexes, and hydride ion abstraction, using $[CPh_3]^+$, from a methylene group incorporated in coordinated cyclic oligoethene complexes can also afford cationic hydrocarbon complexes, *e.g.*, $[Cr(\eta^7\text{-}C_7H_7)(CO)_3]^+$ from the corresponding η^6-cycloheptatriene complex (Section 5.9).

The most important reactions of coordinated alkenes are hydride transfers to generated metal alkyls which lies at the heart of catalytic processes such as hydroformylation (see Section 5.1, p.63).

5.5 Alkyne complexes

Alkynes are generally stronger π acceptors than alkenes, and can bind to transition metals using one of both sets of π bonds, acting as two- or four-electron donors, or as bridging groups. The bonding between alkynes and metals is very similar to that in monoalkene complexes, and some examples of complexes are shown below (Fig. 5.14). In all these compounds, the C≡C bond lengthens, becoming effectively a C=C bond, and in monoalkyne complexes, the alkyne can rotate about the axis from the mid-point of the C–C bond to the metal.

Metal complexes, sometimes even simple salts such as $Ni(CN)_2$, can cause the cyclic oligomerisation of alkynes giving, for example, metal-coordinated cyclobutadienes (Section 5.9), benzenes and cyclooctatetraenes. These reactions are thought to proceed in steps. If metal carbonyls are used, CO is incorporated within the metal-complexed oligomerised alkene, *e.g.*, as in cyclopentadienones, *ortho*-quinones, or tropolones.

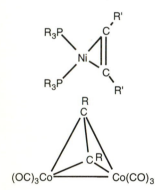

Fig. 5.14 Structures of Ni alkyne and cobalt tetranuclear alkyne carbonyl

5.6 Allyl complexes

Allyl complexes can be prepared by reaction of allyl Grignard reagents with simple metal salts, of allyl halides with metal carbonylate ions and other low oxidation state complexes, or by protonation of coordinated dienes.

The η- or *hapto* nomenclature. If all the C atoms in an unsaturated ligand are coordinated to the metal, the name of the ligand, or its formula, is preceded by η-. The number of C atoms bonded to the metal is indicated by a right-hand superscript, so a delocalised allyl group is designated η^3- and a delocalised cyclopentadienyl group η^5-. Occasionally, σ-bonded ligands are described as η^1-. If there is no superscript number, all C atoms are bonding equally to the metal.

There are two types of allyl ligands: η^1-C_3H_5 (a one-electron donor) which is σ bonded to the metal and has a free C=C bond, and η^3-C_3H_5 (a three-electron donor) which is a delocalised fragment in which all C atoms are interacting with the metal. Allyl ligands can easily undergo $\eta^3 \rightarrow \eta^1$ rearrangements (Scheme 5.5) and this can be assisted by the presence of Lewis bases. For example, addition of phosphine to $[\{Ni(\eta^3\text{-}C_3H_5)(\mu\text{-}Br)\}_2]$ gives first $[Ni(\eta^3\text{-}C_3H_5)(PR_3)Br]$ and then $[Ni(\eta^1\text{-}CH_2CH{=}CH_2)Ni(PR_3)_2Br]$ (ultimately, depending on concentration of PR_3, the allyl bromide is reductively eliminated, giving $[Ni(PR_3)_4]$). Rotation can occur about the Ni–C bond in η^1-allyl–metal groups.

Scheme 5.5 σ/π-interconversion of metal–allyl fragments

Among the most important reactions of metal allyl complexes is the catalytic cyclic di- or tri-merisation of 1,3-dienes. The mechanism of these reactions involves allyl intermediates, and the extent of cyclo-oligomerisation can be regulated by including PR_3. Among the products which can be produced are cyclododecatriene, cycloocta-1,5-diene and 4-vinylcyclohexene.

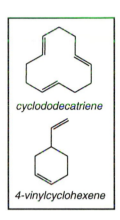

cyclododecatriene

4-vinylcyclohexene

5.7 Cyclopentadienyl complexes

Ferrocene was the first compound to be identified as a 'sandwich' structure with parallel C_5H_5 rings. The coordinated rings did not undergo Diels–Alder addition with dienophiles, indicating that the diene structure was not preserved in the metal complex. However, the rings are susceptible to electrophilic attack analogous to the behaviour of genuine aromatic compounds. This is the origin of the 'ocene' nomenclature and the bis(cyclopentadienyl) metal complexes as a group are known as the metallocenes.

Compounds containing two C_5H_5 rings

Bis(cyclopentadienyl) complexes can be prepared by reaction of the cyclopentadienide ion ($[NR_4]^+$, Li^+ or Na^+ salts) or cyclopentadienyl Grignard agents with anhydrous metal halides or acac complexes.

Only ferrocene has significant air-stability, all other metallocenes readily decomposing on exposure to air. Only cobaltocene can be oxidised to a stable diamagnetic species, $[Co(\eta^5\text{-}C_5H_5)_2]^+$ ($9 + 2{\times}5 - 1 = 18VE$), the cobaltocenium ion, which is isoelectronic with ferrocene.

The electronic structure and hence the magnetic properties of the metallocenes can only be explained by a simple MO description of the complexes which is beyond the scope of this text. The C_5H_5 group provides five MOs all of which participate in bonding with the metal d orbitals. Three donate electrons to the metal, interacting with the metal s, d_{z^2}, d_{xz} and d_{yz} orbitals. The remaining two are empty but interact with the metal d_{xy} and $d_{x^2-y^2}$ orbitals *via* $M \rightarrow C_5H_5$ back-bonding, which is not very significant.

The 'sandwich' structure of ferrocene has long been confirmed crystallographically. The molecule crystallises at temperatures from ambient down to *ca.* 164 K with the two rings staggered, but at lower temperatures the rings are eclipsed. In solution, rotation of the rings occurs about the Fe–

For the purposes of *electron counting*, the $\eta^5\text{-}C_5H_5$ group provides 5 electrons, and so ferrocene obeys the 18-electron rule ($8 + 2{\times}5 = 18$ VE) and is diamagnetic. However, the remaining first-row metallocenes do not: $[V(C_5H_5)_2]$ (15 VE), $[Cr(C_5H_5)_2]$ (*16VE*), $[Mn(C_5H_5)_2]$ (17 VE), $[Co(C_5H_5)_2]$ (19 VE), $[Ni(C_5H_5)_2]$ (20 VE); and these compounds are paramagnetic.

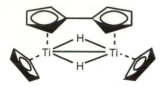

Fig.5.15 'Titanocene'
(4 + 2x5 + 2x1/2 = 1 = 16 VE)

ring midpoint axes. Decamethylferrocene, [Fe(η^5-C$_5$Me$_5$)$_2$] crystallises only in a staggered ring conformation. With the exception of [Mn(C$_5$H$_5$)$_2$], the other metallocenes are presumed to have structures similar to that of ferrocene at room temperature.

Titanocene does not exist as such, attempts to prepare the compounds affording isomers of a compound having the empirical formula [TiC$_{10}$H$_{10}$], one of which is shown in Fig. 5.15. Manganocene has a high-spin d^5 electronic configuration at room temperature but it is close to the point at which it switches from high- to low-spin magnetic behaviour. Its structure at room temperature is ionic and chain-like rather than covalent but, at *ca.* 160 K, it converts to a structure like that of ferrocene. Decamethylmanganocene also has a 'sandwich' structure, which has a magnetic moment, $\mu = 2.18$ BM, consistent with a low-spin configuration.

The rings in ferrocene carry a partial negative charge because, although the C$_5$H$_5^-$ donates charge to the metal, the reduction in charge is partly offset by some back-donation from the metal to the ring. This results in [Fe(η^5-C$_5$H$_5$)(η-C$_5$H$_4$NH$_2$)] being a stronger base than aniline and [Fe(η^5-C$_5$H$_5$)(η-C$_5$H$_4$CO$_2$H)] a weaker acid than benzoic acid. This charge effect also explains why ferrocene is susceptible to electrophilic substitution. Compared to benzene it reacts 3×10^6 times faster. Some reactions are shown in Scheme 5.6. Ferrocene oxidises relatively easily and reversibly to the blue ferrocenium ion, [Fe(η^5-C$_5$H$_5$)$_2$]$^+$. This property can be monitored electrochemically, and is used as the basis of hand-operated devices for the quantitative determination of glucose levels in human blood.

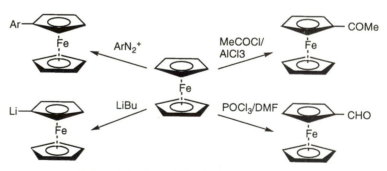

Scheme 5.6 Electrophilic substitutions of ferrocene

The ferrocenyl group can act as an electron donor (ferrocene has a relatively low first ionisation potential), and dipolar molecules such as [Fe(η^5-C$_5$H$_5$)(η-C$_5$H$_4$C$_6$H$_4$CH=CHC$_6$H$_4$CHO)], in which the aldehyde group acts as an acceptor, have important optical properties (*e.g.*, frequency doubling). Decamethylferrocene reacts with tetracyanoethene {(NC)$_2$C=C(CN)$_2$, TCNE} forming charge-transfer salts containing the ions [Fe(η^5-C$_5$Me$_5$)]$^+$ and [TCNE]$^-$, which, as crystalline solids, behave like magnets.

Cobaltocene is quite easily oxidised to the diamagnetic and very stable cobaltocenium ion. This salt is so stable that the methylcyclopentadienyl analogue, [Co(η^5-C$_5$H$_4$Me)$_2$]$^+$, may be converted by nitric acid to the dicarboxylic acid [Co{η^5-C$_5$H$_4$(CO$_2$H)}$_2$]$^+$ without oxidative destruction of

dicarboxylic acid $[Co\{\eta\text{-}C_5H_4(CO_2H)\}_2]^+$ without oxidative destruction of the organometallic core. However, cobaltocene reduces many alkylhalides, generating a cobaltocenium salt and a radical. The radical subsequently adds to unreacted cobaltocene, one η^5-cyclopentadienyl ring being converted to to a η^4-cyclopentadiene (Fig. 5.16), so the net reaction is apparently homolysis of RX and disproportionation of cobaltocene (Co^{II}) to the cobaltocenium ion (Co^{III}) and the cyclopentadienyl(cyclopentadiene) (Co^I) species.

$$[Co(\eta\text{-}C_5H_5)_2] + RX \rightarrow [Co(\eta\text{-}C_5H_5)_2]^+X^- + R\cdot$$

$$[Co(\eta\text{-}C_5H_5)_2] + R\cdot \rightarrow [Co(\eta^5\text{-}C_5H_5)(\eta^4\text{-}C_5H_5R)]$$

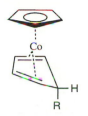

Fig. 5.16. The cobalt (η^5-cyclopentadienyl)(η^4-cyclopentadiene) complex.

Nickelocene is the only metallocene with a $20e^-$ valence electron configuration. It is paramagnetic, and is easily oxidised to $[Ni(\eta\text{-}C_5H_5)_2]^+$ (*19 VE*). One ring of this metallocene is relatively easily reduced to an allylic fragment, affording $[Ni(\eta^5\text{-}C_5H_5)(\eta^3\text{-}C_5H_7)]$ (*18VE*). Protonation of nickelocene rather unexpectedly affords a triple-decker sandwich, $[Ni_2(C_5H_5)_3]^+$, Fig. 5.17.

Although titanocene is a more complicated molecule than its name implies, its derivatives, especially $[Ti(\eta\text{-}C_5H_5)_2Cl_2]$ (*4 + 2x5 + 2x1 = 16VE*), prepared by reaction of $TiCl_4$ with NaC_5H_5, are stable and have a substantial chemistry. The structure of this particular compound is "oyster-like" in that the rings are no longer parallel, being canted at an angle of *ca.* 130° between the mid-points of each ring and the metal. This molecule may be reduced in the presence of π-acceptor ligands to give $[Ti(\eta\text{-}C_5H_5)_2L_2]$ (*18VE*), and the chloride can be nucleophilically displaced by other ligands (Scheme 5.7).

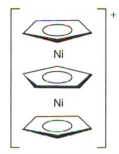

Fig. 5.17. Nickel triple-decker "sandwich".

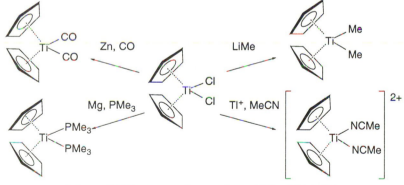

Scheme 5.7. Representatiive reactions of $[Ti(\eta\text{–}C_5H_5)_2Cl_2]$.

Compounds with one C_5H_5 ring

These are sometimes referred to as "half-sandwich" compounds, and are most commonly encountered as mixed carbonyl complexes. They are prepared either by reaction of metal carbonyls or their halides with the cyclopentadienide ion, by reaction of cyclopentadiene (generated by thermolytic cracking of dicylopentadiene) with metal carbonyls (Scheme 5.8), or by treatment of metallocenes with CO under pressure or reaction with $[Ni(CO)_4]$.

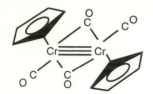

Fig. 5.18 [{Cr(η^5-C$_5$H$_5$)(CO)$_2$}$_2$]

All of the dinuclear species have metal–metal bonds ([{Cr(η^5-C$_5$H$_5$)(CO)$_3$}$_2$]), and two also have bridging CO groups ([{Fe(η^5-C$_5$H$_5$)(CO)$_2$}$_2$] and [{Ni(η^5-C$_5$H$_5$)(CO)}$_2$]). Thermolysis of [{Cr(η^5-C$_5$H$_5$)(CO)$_3$}$_2$] affords [{Cr(η^5-C$_5$H$_5$)(CO)$_2$}$_2$] which contains two terminal and two asymmetrically bridging CO groups as well as a Cr≡Cr bond (Fig. 5.18) (Section 5.11). The species with bridging carbonyl groups are fluxional, the CO groups being exchanged between terminal and bridging positions. In [{Fe(η^5-C$_5$H$_5$)(CO)$_2$}$_2$] this also has the effect of *cis–trans* isomerisation of the rings with respect to the {Fe(μ-CO)$_2$Fe} plane (Scheme 5.9).

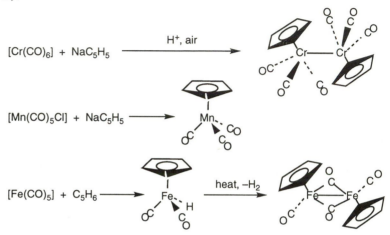

Scheme 5.8 Synthesis of cyclopentadienyl metal carbonyls

These molecules are extremely useful starting materials for a very large variety of organometallic species and in transition-metal-mediated organic syntheses. Some, if not all, of the CO groups may be replaced by other Lewis bases (*e.g.*, alkenes, PR$_3$). Compounds containing M–M bonds may be cleaved by sodium amalgam affording carbonylate ions. Some representative reactions are shown in Scheme 5.10.

Scheme 5.9 CO and ring-site exchange in fluxional [{Fe(η^5-C$_5$H$_5$)(CO)$_2$}$_2$]

[{Fe(η^5-C$_5$H$_5$)(CO)$_2$}$_2$] $\xrightarrow{\text{Na/Hg}}$ Na$^+$[Fe(η^5-C$_5$H$_5$)(CO)$_2$]$^-$

$\downarrow$ X$_2$

RX $\to$ [Fe(η^5-C$_5$H$_5$)(CO)$_2$R]   RCOCl $\to$ [Fe(η^5-C$_5$H$_5$)(CO)$_2${C(O)R}]   H$^+$ $\to$ [Fe(η^5-C$_5$H$_5$)(CO)$_2$H]

[Fe(η^5-C$_5$H$_5$)(CO)$_2$X] $\xrightarrow{\text{CO, X}^-}$ [Fe(η^5-C$_5$H$_5$)(CO)$_3$]$^+$X$^-$

[{Cr(η^5-C$_5$H$_5$)(CO)$_3$}$_2$] $\xrightarrow{\text{NO}}$ [Cr(η^5-C$_5$H$_5$)(CO)$_2$(NO)]

[Mn(η^5-C$_5$H$_5$)(CO)$_3$] $\xrightarrow{\text{NO}^+}$ [Mn(η^5-C$_5$H$_5$)(CO)$_2$(NO)]$^+$

Scheme 5.10 Reactions of cyclopentadienyl metal carbonyls

Like [Mn(CO)$_5$]$^-$ and [Co(CO)$_4$]$^-$, the cyclopentadienyl carbonyl compounds, especially [Fe(η^5-C$_5$H$_5$)(CO)$_2$]$^-$, are powerful nucleophiles.

Of the mononuclear compounds, $[V(\eta^5\text{-}C_5H_5)(CO)_4]$ and $[Mn(\eta^5\text{-}C_5H_5)(CO)_3]$ undergo electrophilic attack, particularly acylation and lithiation, but are less reactive than ferrocene. This reflects the powerful acceptor behaviour of the $\{M(CO)_n\}$ group which substantially reduces the negative charge on the cyclopentadienyl ring.

5.8 Benzene complexes

Bis(benzene)chromium is isoelectronic with ferrocene and has a similar 'sandwich' structure.

Bis(arene) complexes

Bis(arene) metal complexes are prepared by reaction of an anhydrous metal halide with a mixture of Al and $AlCl_3$ in the presence of the arene, usually as solvent, or by co-condensation of metal atom vapour with volatile ligand and solvent at very low temperatures. The latter route is the only way to make Ti complexes and those metal derivatives whose arene ring substituents might be susceptible to reaction with $Al/AlCl_3$ reagents.

The bonding between the benzene ring and a metal is broadly similar to that in cyclopentadienyl compounds. There are six ring MOs of which three donate two electrons each to the metal, the remainder acting as acceptor orbitals. Like its cyclopentadienyl analogue, benzene and its homologues are good donors but relatively poor π acceptors, although $M \rightarrow C_6H_6$ back-donation is more significant than $M \rightarrow C_5H_5$ back-donation. Compared to free benzene, the C–C distance in bis(benzene)chromium is only slightly elongated, and this is consistent with a small amount of back-donation to the ring.

In $[Cr(\eta^6\text{-}C_6H_6)_2]$, the Cr–ring bond energy is 50 kJ mol^{-1} smaller than that in ferrocene (220 kJ mol^{-1}). This is partly due to the arene ring having no charge and therefore there is no electrostatic contribution to the bonding. Consequently, bis(arene) complexes are generally less thermodynamically stable than bis(cyclopentadienyl) complexes. The majority of the bis(arene) complexes are air-sensitive although $[Cr(\eta^6\text{-}C_6H_6)_2]^+$ is air- and water-stable.

Like the bis(cyclopentadienyl) complexes, relatively few of the bis(arene) derivatives obey the 18-electron rule. Bis(benzene)titanium ($4 + 2\times6 = 16$ VE) is diamagnetic, but its vanadium analogue (17 VE) has one unpaired electron and may be reduced to a diamagnetic monoanion (18 VE). $[Cr(\eta^6\text{-}C_6H_6)_2]$ is diamagnetic (18 VE) but its monocation contains one unpaired electron. Whereas $[Fe(\eta^6\text{-}C_6Me_6)_2]^{2+}$ is isoelectronic with bis(arene)chromium, reduction to the neutral bis(arene)iron gives a 20-electron species which is paramagnetic ($\mu = 3.08$ BM, spin-only value for two unpaired electrons 2.83 BM). The related bis(arene)cobalt has a 21-electron configuration, and one unpaired electron. It is possible in neutral bis(arene)metal compounds of the later metals that the rings no longer adopt an η^6-bonding mode but, in solution, may exhibit η^4-interactions (Fig. 5.19).

Bis(arene)chromium derivatives are significantly less subject to electrophilic substitution than ferrocene. This is due to the propensity of the

M

Fig. 5.19 η^4-Arene bonding

compounds to oxidise to $[Cr(\eta^6\text{-arene})_2]^+$ and to the lower ring charge. However one ring may be metallated by $Li(Bu^n)$, so giving access to a small number of ring-substituted complexes.

Compounds containing one C_6H_6 ring

Like cyclopentadienyl compounds, 'half-sandwich' carbonyl compounds, are synthetically useful. These may be prepared by displacement of labile ligands from chromium tricarbonyl derivatives, or by arene ring exchange reactions.

In the chromium complexes, the $Cr(CO)_3$ exerts an electron-withdrawing influence on the arene rings and so, compared with the free arene, the complexed ligand is deactivated towards electrophilic substitution, but becomes more susceptible towards nucleophilic substitution. Thus the rate of substitution of chloride by methoxide in $[Cr(\eta^6\text{-}C_6H_5Cl)(CO)_3]$ is very close to that of *p*-nitrochlorobenzene.

5.10 Other carbocyclic ring complexes

Cyclobutadiene complexes

Free cyclobutadiene can be isolated in low temperature matrices (8–10 K), but it may also be stabilised by coordination to a transition metal fragment. Typical compounds can be prepared by dehalogenation of cyclobutene dihalides, occasionally by dimerisation of acetylenes ($[Co(\eta^5\text{-}C_5H_5)(CO)_2]$ + $Ph\equiv CPh \rightarrow [Co(\eta^5\text{-}C_5H_5)(\eta^4\text{-}C_4Ph_4)]$), or by ligand transfer from palladium complexes such as $[\{Pd(\eta^4\text{-}C_4R_4)X(\mu\text{-}X)\}_2]$.

Free cyclobutadiene is rectangular but when complexed by metals it adopts an η^4-square structure.

The most important cyclobutadiene complex is $[Fe(\eta^4\text{-}C_4H_4)(CO)_3]$ which can undergo aromatic electrophilic substitution (Scheme 5.11). This molecule also serves as a source of free cyclobutadiene which can be liberated by low temperature oxidation. The liberated C_4H_4 may then be trapped by alkynes giving Dewar-benzene derivatives, otherwise difficult to prepare.

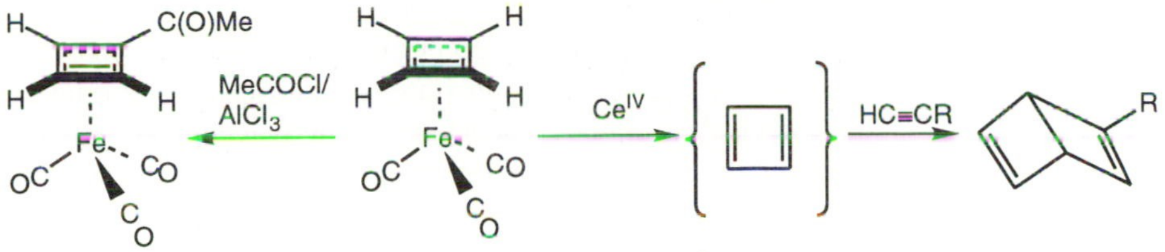

Scheme 5.11 Reactions of $[Fe(\eta^4\text{-}C_4H_4)(CO)_3$

Tropylium or cycloheptatrienyl complexes

Like $C_5H_5^-$ and C_6H_6, $C_7H_7^+$ obeys the *Hückel 4n+2 rule* for aromaticity. The tropylium cation can be isolated as simple salts, *e.g.*, $[C_7H_7][BF_4]$, but these are not useful in the synthesis of organometallic species. Complexes of this ring (a seven-electron donor) may be generated by direct reaction of cycloheptatriene with metal complexes or by hydride abstraction from cycloheptatriene metal carbonyls.

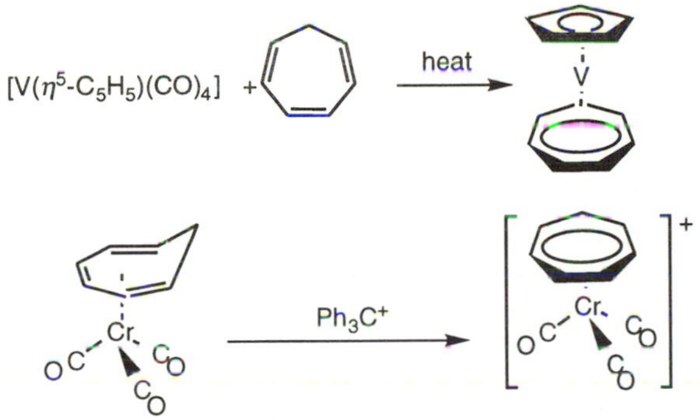

The C_7H_7 ring in tropylium complexes is planar and the C–C bond lengths are equal. The cationic derivatives are susceptible to nucleophilic addition, regenerating cycloheptatriene complexes.

Cyclooctatetraene complexes

Free cyclooctatetraene, C_8H_8, does not obey the Hückel aromaticity rule, but may complex to metal fragments as an η^4-dienoid {to $Fe(CO)_3$} or η^6-trienoid {to $Cr(CO)_3$} group, or even as a chelating diolefin (see Scheme 5.4).

However, the $[C_8H_8]^{2-}$ dianion is a 10π-electron system and is therefore aromatic and planar. This bonding mode is rare in first-row transition metal chemistry, being found only in titanium chemistry. Reaction of $[Ti(OBu)_4]$ (Section 4.1) with C_8H_8 and $AlEt_3$ affords $[Ti(\eta^8\text{-}C_8H_8)(\eta^4\text{-}C_8H_8)]$, in which one ring is planar and the other binds *via* a *cis-s*-η^4-diene interaction, and $[Ti_2(C_8H_8)_3]$, which has a 'triple-decker sandwich-like' structure where two $[Ti(\eta^8\text{-}C_8H_8)\}$ groups are bridged by a non-planar C_8H_8 ring. Reduction of

$[Ti_2(C_8H_8)_3]$ gives $[Ti_2(C_8H_8)_3]^{2-}$ in which the 'triple-decker' structure is retained but the bridging C_8H_8 ring is now fully planar.

5.11 Copper(I)

Copper has a single s electron outside the filled $3d$ shell and it is for this reason that the +1 oxidation state is significant. The relative stabilities of Cu^I and Cu^{II} in aqueous media are strongly dependent on concentration and, particularly, the anions and ligands present (Section 3.2). The equilibrium

$$2\ Cu^I \ \rightleftharpoons \ Cu^0 \ + \ Cu^{II}$$

is pushed to the left by polarisable ligands, or those capable of π bonding, such as CN^- or thioethers. With anions which are incapable of covalent bonding or forming bridges, or with ligands which form particularly stable complexes with Cu^{II}, the equilibrium shifts to the right. While the stability of Cu^+ in water is low, in acetonitrile it is high, and salts of $[Cu(NCMe)_4]^+$ can be isolated and are useful intermediates in the synthesis of Cu^I complexes.

Copper(I) complexes with halide and other simple ligands have a wide variety of structures ranging from mononuclear to dinuclear chains to tetranuclear structures. While the copper coordination can be linear or trigonal-planar, tetrahedral geometry is preferred, in contrast to Cu^{II}, which can be four-, five- or six-coordinate. This has important structural consequences for the relative stabilities of Cu^I and Cu^{II}. In a redox reaction involving the Cu^{II}/Cu^I couple, the stability and accessibility of either one of these oxidation states will depend crucially on the ability of the ligand system to adapt to the preferred coordination and geometry of the metal ion.

Alkyl and aryl compounds are prepared by reaction of copper(I) halides with organolithium or Grignard reagents. These may be polymeric, like $[\{CuMe\}_n]$, or oligonuclear, perhaps with ring structures, like $[\{Cu(mes)\}_5]$ (mes = mesityl). The most important organo-copper compounds are the lithium alkyl cuprates, $LiCuR_2$, made by reaction of CuI or CuMe with LiR. Some organo-cuprate ions have linear C–Cu–C structures.

Copper(I) forms alkene and alkyne complexes in which the bonding from the hydrocarbon to the metal is predominantly σ donation from the π-bonding orbital to an acceptor orbital on the copper. Copper(I) ammine complexes react with monoacetylenes, RC≡CH, to give oligonuclear acetylides in which one Cu–C≡CR group is π bonded to another. These are important reagents in the synthesis of a variety of organic acetylenic compounds by reaction with aryls and other halides, and may be key intermediates in the oxidative dimerisation of monoacetylenes:

$$2\ RC{\equiv}CH + 2Cu^{2+} + 2py \ \rightarrow \ RC{\equiv}C{-}C{\equiv}CR + 2Cu^+ + 2[pyH]^+$$

5.12 Metal–metal bonding

At the beginning of this chapter the structures of di-, tri- and other polynuclear metal carbonyl compounds were illustrated, and many of these

Copper(I) oxide, Cu_2O, is a red pigment used in some paints, glasses, porcelains, glazes and ceramics. The analogous sulphide, Cu_2S, is black and insoluble, and is used in solar cells, electrodes and, occasionally, as a solid lubricant. Copper alkoxides, with the exception of the insoluble polymeric methoxide, are oligonuclear but are useful reagents for metallating acidic hydrocarbons and for converting alkyl halides into ethers.

Lithium alkyl cuprates are useful in C–C bond formation, especially in reaction with organic halides, and in 1,4-additions to α,β-unsaturated ketones to give saturated alkylated ketones.

Copper (I) can form CO complexes, *e.g.*, $[Cu(CO(bipy)]^+$, but these are not as stable as the earlier transition metal carbonyls.

compounds were shown to have metal–metal bonds (Fig. 5.1). The formation of metal–metal bonding is almost a prerequisite of low oxidation state chemistry. If manganese carbonyl was monomeric, it would be paramagnetic, so the occurrence of diamagnetism in a system in which the simplest empirical formula would imply that the species should be paramagnetic is a strong suggestion that metal–metal bonds are present. It is not possible *a priori* to determine whether the compound will be dinuclear or will have higher nuclearity (as in a cluster), nor whether the metal–metal bond will be the only connection between the empirical fragments, or whether there will be other bridging groups, such as CO in [$Fe_2(CO)_9$]. The most clear indication of metal–metal bonding is, of course, the intermetallic distance as determined by X-ray diffraction methods.

While the compounds in Fig. 5.1 have M–M bond orders of one, it is possible to have higher orders. For example, thermolysis of [{$Cr(\eta^5\text{-}C_5H_5)(CO)_3$}$_2$] (each metal atom has 18 VE) gives [{$Cr(\eta^5\text{-}C_5H_5)(CO)_2$}$_2$] (Fig. 5.18) in which there is a Cr≡Cr triple bond (3 VE are apportioned to each metal for the metal–metal bond). The effect of bond multiplicity is clear from the Fe–Fe bond lengths in [{$Fe(\eta^5\text{-}C_5H_5)(CO)(\mu\text{-}CO)$}$_2$] (Fe–Fe), 2.49 Å and [{$Fe(\eta^5\text{-}C_5H_5)(\mu\text{-}NO)$}$_2$] (Fe=Fe), 2.33 Å: there is a contraction as bond multiplicity increases.

However, metal–metal bonding can occur in systems other than metal carbonyls. Examples include [$Co_2(CN)_{10}$]$^{6-}$, which contains Co^{II} and has the [$Mn_2(CO)_{10}$] structure, and a series of dinuclear species containing divalent metal ions linked by carboxylato, triazenido ($RNNNR^-$) and analogous bridging ligands in which two donor atoms are separated by one other atom. The occurrence of metal–metal bonding is best illustrated by [$Cr_2(OAc)_4L_2$] and its copper(II) analogue. The chromium complex is very unusual in that it is diamagnetic, since Cr^{II} has a d^4 electronic configuration and mononuclear species must be paramagnetic. However, the dinuclear structure was established by X-ray crystallography, and the Cr–Cr distance varies from 2.3–2.5 Å, depending on L (Fig. 5.20). In [$Cr_2(OAc)_4$] itself, the Cr-Cr separation is extremely short, 1.96 Å, commensurate with a quadruple bond, and in complexes containing other types of bridging ligand it can be as low as 1.83 Å. In dinuclear copper(II) carboxylates and related triazenido complexes, the Cu–Cu bond varies between 2.7 to 2.8 Å, which is rather too long for significant metal–metal bonding.

The quadruple bond may be explained if it is assumed that within each planar CrO_4 unit, the Cr–O bonds define the x and y axes and the Cr–Cr bond lies on the z axis. Then the $d_{x^2-y^2}$ orbital on each metal is exclusively occupied in σ bonding with the ligands (Fig. 5.21). However, the d_{z^2} orbitals are available for metal–metal σ bonding, the d_{xz} and d_{yz} orbitals for metal–metal π bonding, and the d_{xy} for weak δ bonding: hence the maximum possible bond order of four.

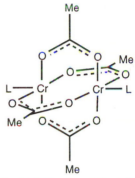

Fig. 5.20 Structure of [$Cr_2(OAc)_4L_2$]

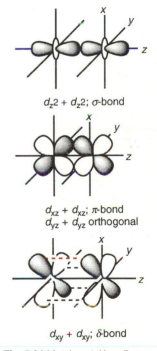

$d_{z^2} + d_{z^2}$; σ-bond

$d_{xz} + d_{xz}$; π-bond
$d_{yz} + d_{yz}$ orthogonal

$d_{xy} + d_{xy}$; δ-bond

Fig. 5.21 Metal–metal bonding

6   Bio-transition metal chemistry

For more detailed accounts of bio-transition metal chemistry, consult D. E. Fenton, 'Biocoordination chemistry', Oxford Chemistry Primers 25, 1995, and P. C. and R. G. Wilkins, 'Inorganic chemistry in biology', Oxford Chemistry Primers 46, 1997.

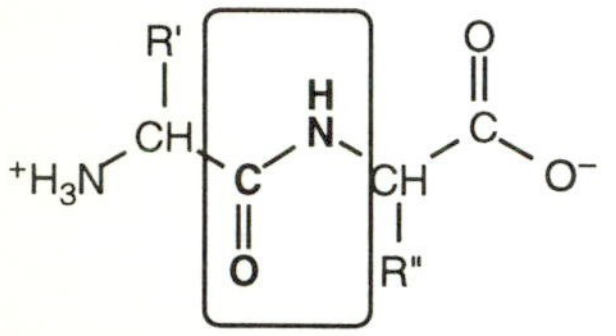

the peptide link

Fig. 6.1 Basic structure of a peptide or amido link

R = H   glycine (gly)
R = CH_2SH cysteine (cys)
R = CH_2CH_2SMe methionine(met)
R = CH_2CO_2H aspartic acid
R = $CH_2CH_2CO_2H$ glutamic acid

R = H_2C ⟨ ⟩ OH tyrosine

R = H_2C ⟨ ⟩ histidine (his)

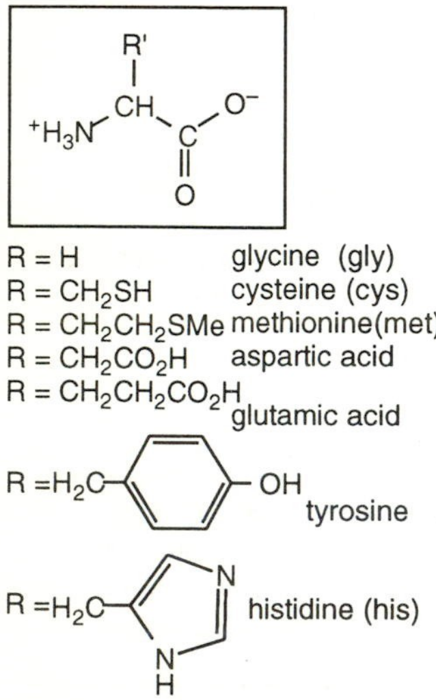

Fig. 6.2 Naturally occurring amino acids

It appears that 13 metals are essential for plants and animals, of which four, Na, K, Mg and Ca, the so-called *bulk metals,* are present in relatively large amounts. The other nine are the so-called *trace metals*, which includes V → Cu, Zn and Mo. Of this group, iron, copper and zinc are at the top end of the scale; the other metals are present in extremely low concentrations. There is, as yet, no detectable role for Ti in biochemical processes.

Iron and copper are essential for most, if not all, life forms. These metals are characterised by readily interchangeable oxidations states: Fe^{III}/Fe^{II} and Cu^{II}/Cu^{I}, and so are important in electron-transfer systems, *i.e.*, redox reactions. They are also involved in oxygen-storage and -transport. The other $3d$ block metals which have so far been identified as active in life processes occur in enzymes, Nature's catalysts, and serve a variety of purposes.

The functions of metals in biology may be generally classified into two major groups: metalloproteins, which are further sub-divided into transport and storage functions, and metalloenzymes; and non-protein systems, also involved in storage and transport, and in photo-redox activity.

Proteins consist of one or more polypeptide chains. Polypeptide chains are comprised of amino acids usually connected by a peptide (or amido) bond (Fig. 6.1). Some commonly encountered naturally occurring amino acids are shown in Fig. 6.2.

The *metalloproteins* incorporate one or more metal atoms as a normal part of the tertiary structure of the protein. *Metalloenzymes* are a sub-class of this group, and the term is applied to systems which not only require the participation of metal ions at the active site to function, but also bind the metal ion or ions strongly, even in the resting (non-functioning) condition. In a number of enzymes which are structurally characterised, the metal ion is bound in an unusual, perhaps enforced, stereochemistry, known as the *entatic state*, which appears to enhance its ability to bind and/or activate the substrate.

It is possible in some cases to remove the metal ions and then either restore them or replace them with other ions without serious disruption of the protein structure. The demetallated system is known as an *apoprotein*, which implies that the metalloprotein function is virtually totally recovered upon reconstitution with the naturally occurring metal ion.

In this chapter, many of the rôles of first-row transition metals in biological chemistry are outlined. The examples chosen are either those where a structure of at least one metalloprotein has been determined crytallographically, or where there are appropriate model complexes which mimic biological function realistically.

6.1 Transport and storage

In the following sections, particular mention is made of transport and storage of metals, especially Fe, electron transfer, and transport and storage of dioxygen.

Transport

Transition metal ions must be mobilised from their point of uptake from the environment, protected and delivered to the organs which require them, and stored, if there is an excess, until the metal is required. Part of the transport phenomenon requires passage of these metal cations across biological barriers, *i.e.*, cell membranes. The $3d$ metals usually precipitate, as hydrated oxides, at biological pH values and are frequently toxic since they can generate small amounts of radicals, *e.g.*, O_2^- or $HO^{\bullet}$, which can attack and rupture the amino acid constants of proteins, thereby destroying the functions of metalloproteins and enzymes. To deal with these possibilities, the metal ion must be complexed by appropriate carrier/protective ligands.

Iron is probably the best understood metal in this respect, and its transportation modes differ in vertebrates (*e.g.*, humans) and in bacteria. In the former, iron is extremely efficiently recycled and very little is absorbed from diet or is excreted under normal conditions. Only about 1% of iron in the body is used at any one time (*e.g.*, in respiration or electron transfer processes) and the rest is stored. The transport proteins (*transferrins*) include serum transferrin and lactoferrin which is present in milk. These are monomeric species with molecular weights of *ca.* 80000) which bind two Fe^{III} ions tightly but reversibly in two separated but linked domains. The formation or stability constant for the metal ions is in excess of 10^{20} M^{-1}. The iron site in each domain appears to be six-coordinate, there being two phenolate ligands (tyrosine), one N (histidine), a monodentate carboxylate (aspartate) and a bidentate carbonate ion. The precise modes of action of transferrins are not fully understood but among their roles is protection of the body against invading microbes which require iron, presumably by collecting excess iron and delivering it to a storage protein.

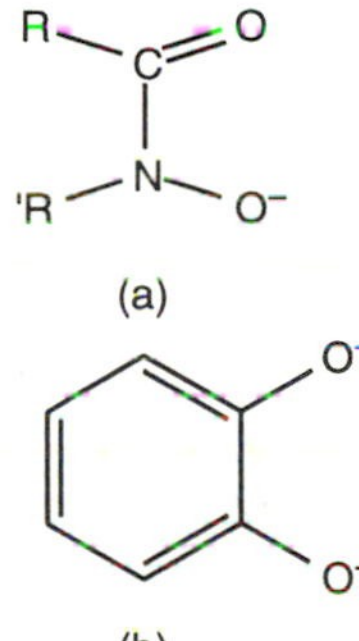

Most micro-organisms require iron for growth, and in bacteria iron is mobilised by *non-protein* chelating ligands known as *siderophores*. These are widespread in nature, and are derived either from hydroxamate (*ferrichromes, ferrioxamines*) or catecholate (*enterobactins*) ligands (Fig. 6.3). These ligands provide O donor atoms to stabilise Fe^{III} with stability constants in excess of 10^{30} M^{-1} (a biological example of high stability associated with chelation and hard acid/hard base pairing). The Fe^{III} centres are high-spin and are kinetically labile which, of course, is important to the extraction and exchange of iron. This property also permits replacement of iron by other metals, including the kinetically stable Cr^{III}, which allows detailed spectroscopic studies of the siderophore ligand superstructures, and by V^{IV} which has facilitated X-ray crystallographic studies of an enterobactin complex. It appears that the mode of release of iron from siderophores involves reduction of the Fe^{III} species to Fe^{II} whose stability constants are significantly lowered (*ca.* 10 M^{-1}).

(a)

(b)

Fig. 6.3 (a) Hydroxamate and (b) catecholate ligands

A demetallated siderophore, *desferrioxamine B*, has been used effectively in the treatment of iron overload diseases (p.90).

Metallothioneins are proteins rich in cysteine residues which can act as copper ion carriers, possibly as Cu^I.

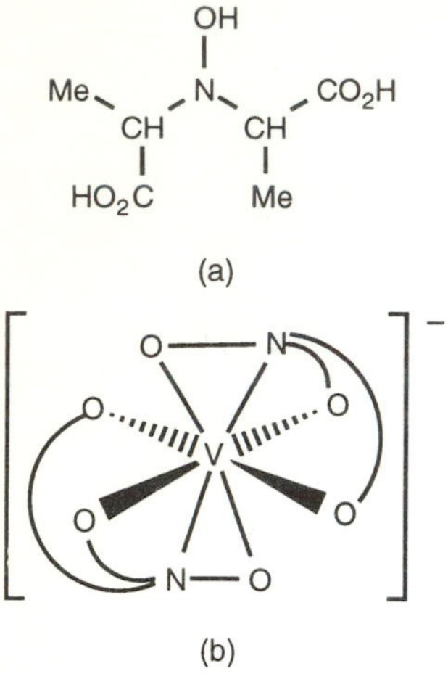

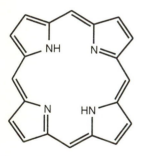

(a)

$$[\quad]^-$$

(b)

Fig. 6.4 (a) HIPDAH$_3$, and (b) its coordination to vanadium in [V(HIPDA)$_2$]$^-$

The *cytochromes* contain an Fe porphyrinato (2–) core and are *haem* proteins. Ferredoxins do not contain porphyrinato ligands; they are *non-haem* proteins.

Fig. 6.5 The basic ligand in haem proteins

Vanadium is extracted from soil and accumulated by a mushroom (*Amanita muscaria*). The metal, as V^{IV}, is extracted as a low molecular weight complex, *amavadin*, which contains 2,2'-(hydroximino)dipropionic acid [HIPDAH$_3$], Fig. 6.4(a), as a ligand. This complex may be oxidised to a V^V species, [V(HIPDA)$_2$]$^-$, whose structure is known. The metal is eight-coordinate (Fig. 6.4(b)), each ligand providing two monodentate carboxylato groups and a very unusual η^2-N–O interaction, similar to the bonding in some hydroxylamine species (Chapter 4, p.56). It is thought that amavadin is involved in electron transfer processes.

Electron transfer proteins

Among this class of biomolecule are the cytochromes, iron–sulphur cluster proteins and blue copper proteins. Almost all biological electron transfers reactions occur *via* an outer-sphere mechanism (Section 3.4).

The *cytochromes* are haem proteins (see Fig. 6.5) in which the iron centre is coordinated by a planar porphyrinate ring and may carry one or two additional axial ligands. These molecules are present in all forms of aerobic life: in plants where they are involved in photosynthesis, and in a very wide range of cells where they are intimately involved in the process of respiration.

Cytochromes are classified according to the substituents on the porphyrinate ring, whether the haem group is covalently bound to the protein or not, whether the ring is partially reduced, whether the metal is four- or five-coordinate, and on the type of axially coordinated groups whose ligand field strength controls the spin-state of the iron. Because the majority of cytochromes contain six-coordinate iron, they cannot bind small molecules and so are exclusively involved in one-electron transfer, interconverting between the Fe^{II} and Fe^{III} forms. Several cytochromes carry axial histidinyl (N) and methioninyl (neutral S) ligands which are attached to the protein backbone. From structural studies of a particular oxidised and reduced cytochrome it is clear that electron transfer occurs over quite long distances (>16 Å), but without significant conformation changes.

Iron–sulphur electron transfer proteins are encountered in all living organisms and are known collectively as *ferredoxins*. There are a number of structural types: mononuclear species containing Fe coordinated by four S (cysteinyl) atoms, or clusters containing Fe$_2$S$_2$-, Fe$_3$S$_4$- or Fe$_4$S$_4$- groups. In the clusters, each Fe atom is coordinated by four sulphur atoms, either 'inorganic' S^{2-} or protein-bound cysteinyl S$^-$.

The simplest iron–sulphur protein is *rubredoxin* which is found in bacteria. It contains one Fe atom coordinated by the S atoms of four cysteinyl residues and, as it has a distorted tetrahedral geometry, it may be considered to be in an entatic state. Rubredoxins appear to act as one-electron donor–acceptor proteins, which involves shuttling between high-spin Fe^{II}, [Fe(S-cys)$_4$]$^{2-}$, and high-spin Fe^{III}, [Fe(S-cys)$_4$]$^-$, states.

The di-iron disulphide proteins, represented as [Fe$_2$S$_2$(S-cys)$_4$]$^{n-}$ (n = 2 or 3), occur in plants and bacteria where they are associated with photosynthesis, and in mammals. The structure of the iron sulphur core is based on two tetrahedra sharing a common edge (Fig. 6.6(a)). This dinuclear system is a one-electron donor or acceptor:

$$[Fe_2S_2(S\text{-cys})_4]^{2-} \rightleftharpoons [Fe_2S_2(S\text{-cys})_4]^{3-}$$

in which the oxidised form contains two high spin d^5 Fe^{III} centres which are strongly coupled magnetically making the dinuclear species diamagnetic. One electron reduction generates a mixed-valence $Fe^{II}Fe^{III}$ species in which the Fe^{II} and Fe^{III} centres retain their identity (there is no delocalisation between the metals).

The $[Fe_4S_4(S\text{-cys})_4]^{n-}$ clusters have slightly distorted cubane-like structures (Fig. 6.6(b)) and constitute a three-membered electron transfer series:

$$[Fe_4S_4(S\text{-cys})_4]^{3-} \rightleftharpoons [Fe_4S_4(S\text{-cys})_4]^{2-} \rightleftharpoons [Fe_4S_4(S\text{-cys})_4]^{-}$$

The most reduced (trianion) and most oxidised (monoanion) forms contain paramagnetic delocalised $\{Fe^{II}_3Fe^{III}\}$ and $\{Fe^{II}Fe^{III}_3\}$ cores, whereas the dianion is diamagnetic, having a $\{Fe^{II}_2Fe^{III}_2\}$ core. Although this cluster system has three oxidation levels, only one pair is used biologically, depending on the source of the protein. The potential range for electron transfer is nearly 0.75 V, but it appears that it is the protein rather than the cluster which dictates the particular redox pattern chosen by the natural system. Models of these iron–sulphur proteins have been made (Scheme 6.1).

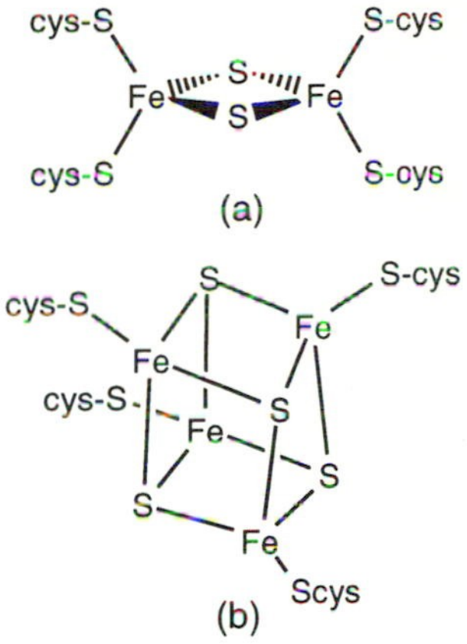

(a)

(b)

Fig. 6.6 Iron–sulphur centres in ferredoxins: (a)Fe$_2$S$_2$; (b)Fe$_4$S$_4$

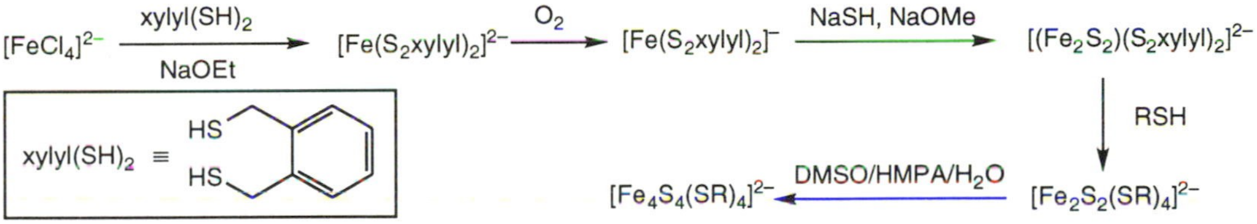

Scheme 6.1 Synthesis of iron–sulphur clusters

The *blue copper proteins* are so called because of an intense absorption near 600 nm ($\varepsilon > 3000$ M^{-1} cm^{-1}) caused by S(cysteinyl) $\rightarrow$ Cu^{II} charge transfer. *Plastocyanin*, which occurs in higher plants and green algae, is involved in electron transfer in photosynthesis, and *azurin*, found in denitrifying bacteria, facilitates electron transfer in some respiratory processes. Both proteins exploit the Cu^{II}/Cu^{I} redox couple which is greatly influenced by the stereochemical requirements of the different oxidation states: tetrahedral for Cu(I) and tetragonally distorted octahedral for Cu(II). X-Ray crystallographic studies of both proteins in both oxidation states has shown that there is very little change in gross structure on reduction from the Cu(II) to the Cu(I) states. In plastocyanin, the Cu site has a distorted tetrahedral geometry with an unusually long Cu–S(methionine) bond, regardless of formal metal oxidation state (Fig. 6.7(a)). In azurin, there is a slight expansion in the copper–ligand bond lengths on reduction, as expected by the increase in ionic radius. The metal has distorted trigonal prismatic geometry (Fig. 6.7(b)).

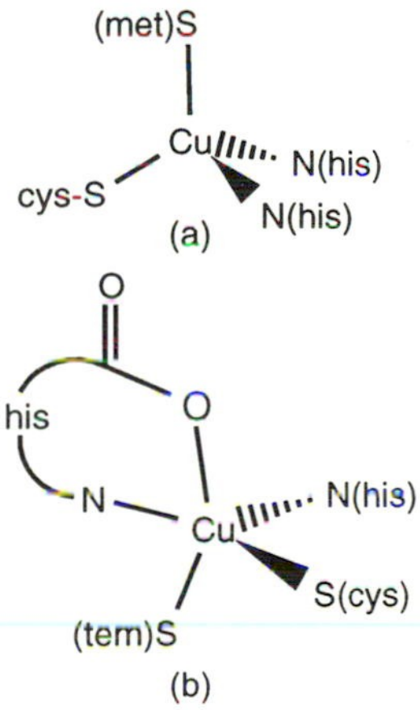

Fig. 6.7 Metal sites in (a) plastocyanin and (b) azurin

Oxygen transport

Proteins which are responsible for the transport and storage of dioxygen are sometimes referred to as the 'respiratory pigments'. These are required by all animals and many other species. The most common respiratory pigment in man is, of course, *haemoglobin*, an iron-based species which has a deep red colour the intensity of which depends on whether the protein is oxygenated or not. However, not all species use haem proteins. Many invertebrates (lobsters and crabs, squid, octopus, snails and marine worms) employing *haemocyanin*, a dinuclear copper species which gives blood a blue colour, or *haemerythrin* which uses two iron atoms and imparts a violet colour to blood. The prefix 'haem' is confusing, as it is often used in conjunction with proteins containing an iron porphyrinate ring, but of the respiratory pigments, only haemoglobin contains this group.

Haemoglobin has a molecular weight of *ca.* 65 000 and is tetrameric. *Myoglobin* can be regarded as its monomeric analogue. The haem unit is a Fe^{II} protoporphyrin IX group (Fig. 6.8(a)).

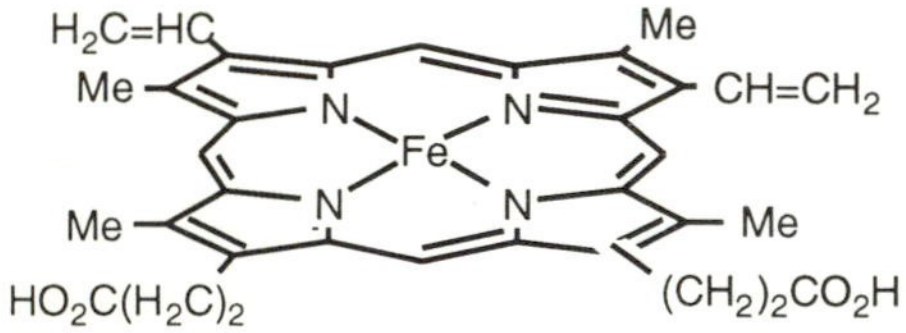

Fig. 6.8 (a) Fe^{II} protoporphyrin IX, and (b) a 'picket fence' porphyrin

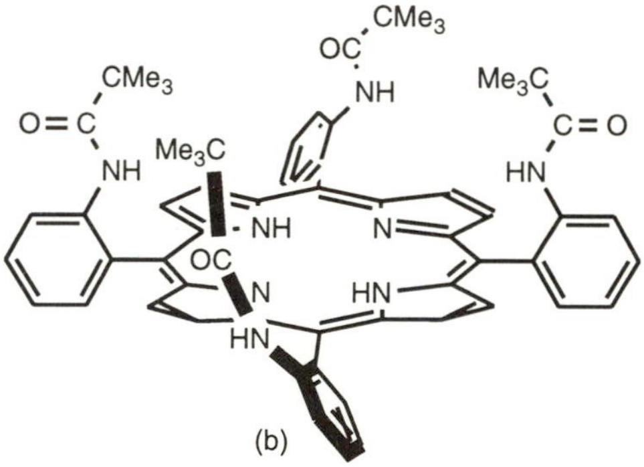

These species absorb one molecule of O_2 per iron atom, the dioxygen ligand being monodentate with an Fe–O–O bond angle varying between 115° (*oxy*myoglobin) and 156° (*oxy*haemoglobin). In the *deoxy* forms the iron atom is nearly square-pyramidal with an axial histidine N atom, the high-spin Fe^{II} atom lying above the plane of the porphyrinate ring. Upon O_2 absorption, the metal becomes six-coordinate. A considerable amount of work on model dioxygen complexes has revealed a lot about the natural oxygen-carriers. It is clear that the iron centre in porphyrinate complexes must be protected sterically, since reaction of simple planar porphyrinato complexes leads to inactive dinuclear species containing an Fe–O–Fe linkage, *e.g.*, [L(porph)Fe–O–Fe(porph)L] (porph = porphyrinate) where L is a Lewis base ligand. However, the porphyrin ring can be modified by attaching substituents which constitute a kind of 'picket fence' around the iron centre (Fig. 6.8(b)), which not only prevents dimerisation, but also provides a hydrophobic pocket at the iron site for interaction with O_2 in an end-on fashion as Fe^{III}–O_2^-.

Three terms occur frequently in the context of oxygen-carrying proteins: *deoxy, oxy* and *met*. The first refers to proteins in which the metal is fully reduced. The second applies to the oxygenated product formed by reaction of O_2 with a deoxy form, and the third is a protein in which the metal centre is oxidised and unreactive towards O_2.

The reactions of cobalt(II) complexes with O_2 to form peroxide and superoxide complexes were mentioned in Section 4.1. Only a few such complexes can function as O_2-carriers and one is shown in Scheme 6.2.

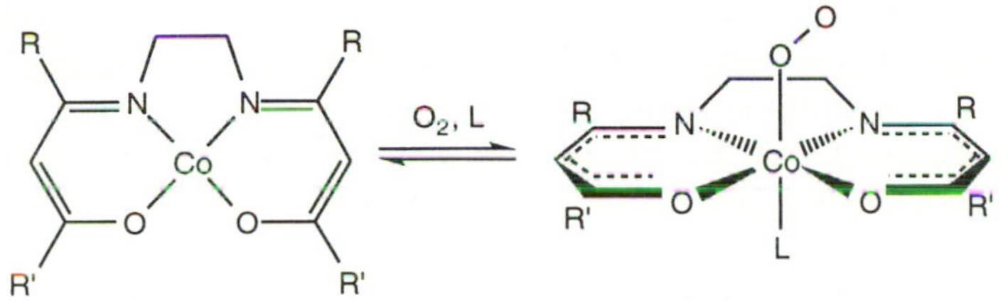

Scheme 6.2. Cobalt–oxygen complexes.

The structure of one complex (R = Me, R' = Ph, L = DMSO) showed that the Co–O–O bond angle is 125° and the O–O distance 1.26 Å, close to that of the superoxide ion. So the cobalt–dioxygen interaction in these complexes is similar to that in iron 'picket fence' and related porphyrins.

Haemerythrins are dinuclear iron proteins whose active sites consist of two Fe atoms triply bridged by an O atom and two carboxylato groups associated with amino acids. The remaining five ligands are histidines, so that one Fe atom is six- and the other five-coordinate in the *deoxy* form. It also seems likely that the oxo bridge is protonated (*i.e.*, OH) in the *deoxy* form, the five-coordinate iron is present as Fe^{II} and the intermetal distance is 3.3 Å. Dioxygen binds to this iron centre giving a $\{Fe^{III}(\mu\text{-}X)(\mu\text{-}Y)_2Fe^{III}\text{-}O_2^{2-}\}$ group in the *oxy* form. A possible mechanism for the uptake and release of O_2 from haemerythrins is shown in Scheme 6.3.

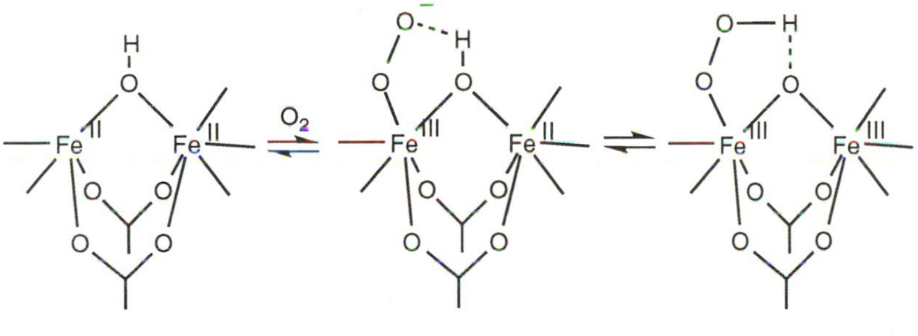

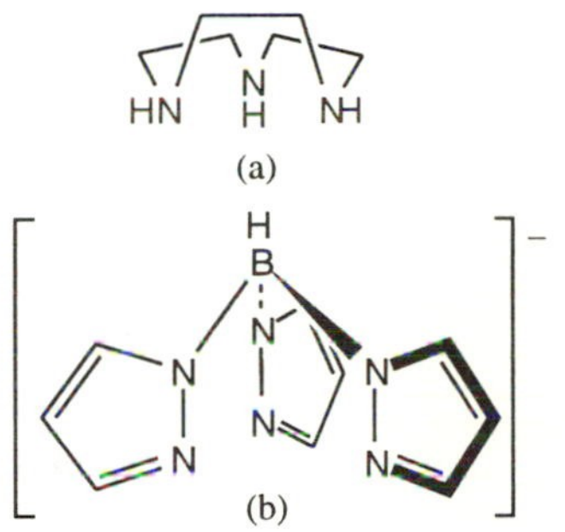

Fig. 6.9 (a) tacn, and (b) Tp⁻ ligands

Scheme 6.3 O_2 uptake and release by haemerthyrins: a possible mechanism

Met-haemerythrin contains two six-coordinate Fe^{III} centres and is inactive towards O_2. Attempts to make models for haemerythrins usually lead to species very similar to the *met* form, *viz*. [LFe(μ-O)(μ-OAc)$_2$FeL] where L is a tridentate capping N-donor ligand, tacn or Tp⁻ (Fig. 6.9).

Haemocyanins are dinuclear copper proteins, the *deoxy* form containing two Cu^I centres which are coordinated by three histidinyl groups (N). From model studies of copper chemistry with sterically hindered Tp analogues, it has been established that in the *oxy* form each metal retains the three histidinyl groups but are symmetrically bridged by an O_2^{2-} ligand (Fig. 6.10). This was subsequently confirmed by a crystallographic study of an *oxy*haemocyanin obained from horseshoe crabs.

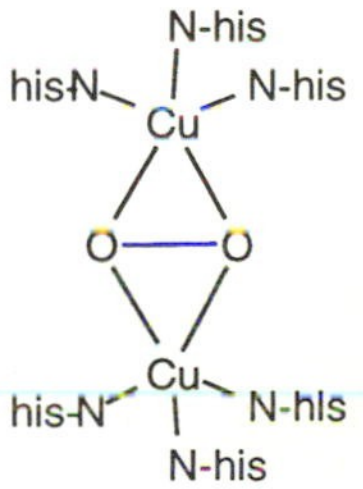

Fig. 6.10 Model haemocyanin species

Storage

Iron is stored in animals, many plants, and some bacteria, mainly in a 'bio-mineral' *ferritin*. Ferritins are a group of large, approximately spherical proteins (diameter *ca.* 130 Å) encapsulating hydrated Fe^{III} oxide quite similar to a particle of rust (Chapter 4). This mineral core, which may contain up to 4500 iron atoms, has an approximate composition $[\{FeO(OH)\}_8\{FeO(PO_4H_2)\}]$, with a structure akin to a close-packed array of oxide and hydroxide ions, the Fe atoms occupying octahedral 'holes'. The phosphate residues may anchor the iron oxide particles to the surrounding protein shell. Hydrophilic and hydrophobic channels connect the inner mineral core to the outer environment, enabling iron and other agents to enter and leave. The mode of core formation and extrusion of iron is not fully understood, but it appears that the core can only be formed from aqueous Fe^{II} and that oxidation to Fe^{III} only occurs inside the protein casing. Potential model cluster oxy-iron species, *e.g.*, $[Fe_{11}O_6(OH)_6(O_2CPh)_{15}]$, can be assembled by slow reaction of salts of $[Fe_2OCl_6]^{2-}$ with benzoate ion in acetonitrile. The 11 iron atoms are at the vertices of a distorted penta-capped trigonal prism with 12 triply bridging O atoms and 15 doubly bridging benzoate groups, each iron having distorted octahedral coordination.

6.2 Metalloenzymes

Enzymes are catalysts and metalloenzymes comprise about 40% of all enzymes. First-row transition metals, especially Fe and Cu but also Mn and Ni, play the crucial role in a group of metalloenzymes known as the *oxidoreductases*, which catalyse a wide variety of biological oxidations and reductions. Oxidases use O_2, superoxide and peroxide ion as oxidants in aerobic organisms, and sulphate and other oxidants in anaerobic systems. The hydrogenases use H_2 as reductant.

Mono-oxygenases catalyse the insertion of one O atom of dioxygen into a substrate, *i.e.*, conversion of C–H to C–OH, the other O being reduced to water. A most important mono-oxygenase is the haem-containing enzyme *cytochrome P-450*. This enzyme is important in the body's defences against hydrophobic compounds such as drugs and pesticides, converting them to water soluble and therefore easily excreted compounds. The structure and catalytic behaviour of this species is reasonably well understood, the haem fragment shuttling between five- and six-coordinate species, the metal centre existing in formal oxidation states II, III and IV (as the ferryl group FeO^{2+}). The proposed catalytic cycle is shown in Scheme 6.4.

The 'resting condition' of the enzyme contains high-spin Fe^{III}, the axial ligands being H_2O and S^- (cysteine). The hydrocarbon substrate then binds in the proximity of the haem group (there is no evidence to suggest that the substrate actually binds to iron), which retains its high-spin configuration but becomes square pyramidal. Electron transfer causes reduction to pyramidal high-spin Fe^{II}, and O_2 then binds, generating initially a six-coordinate low-spin Fe^{III} superoxo complex. Further reduction affords a low-spin Fe^{III} peroxo complex which, following proton transfer, releases H_2O with generation of a low-spin ferryl species ($Fe^{IV}O$), which inserts the O atom

> Note that *cytochrome P-450*, so-called because of an electronic absorption present in its CO adduct, is an O-transfer enzyme: most other cytochromes, *e.g.*, *cytochrome c*, are electron-transfer proteins.

into the hydrocarbon substrate. An important competing reaction of the high-spin five-coordinate Fe^{II} precursor is its reaction with CO to irreversibly generate a highly stable monocarbonyl complex. This reaction, common to all five- or potentially five-coordinate Fe^{II} haem compounds, effectively blocks the oxygen-transport and -activation functions of these compounds, and has catastrophic consequences for species whose life depends on these functions. Nitrogen monoxide, NO, behaves similarly.

Mono-oxygenases and model compounds have been used successfully to catalyse epoxidation of a variety of alkenes.

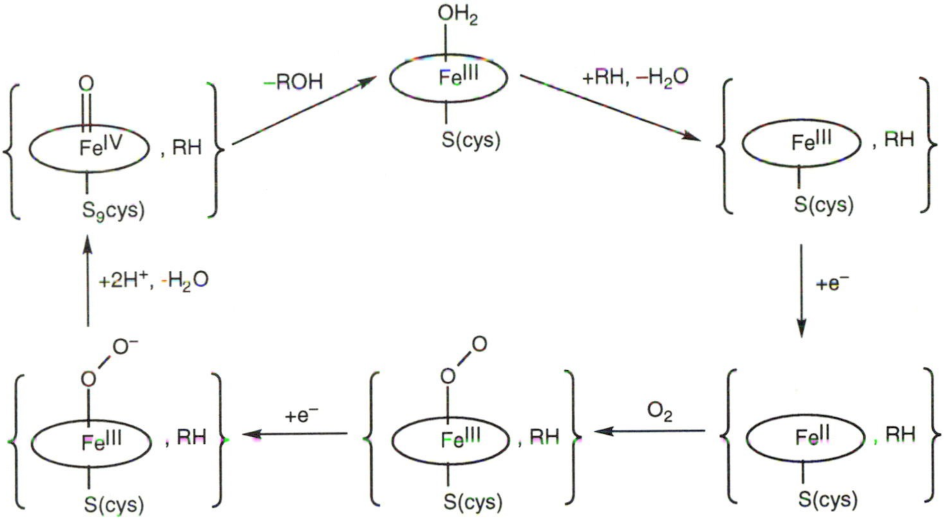

Scheme 6.4 Proposed catalytic cycle for cytochrome P-450.

Methane mono-oxygenase catalytically converts CH_4 to methanol. One form of this enzyme contains a di-iron centre, and the structure of an extract reveals a coordination arrangement somewhat reminiscent of that in haemerythrins. The iron atoms, which have distorted octahedral geometries, are bridged by OH^-, a carboxylato group of a glutamate residue, and an acetate (Fig. 6.11). This last group may be derived from the crystallisation of the compound, however, so may not be integral to the structure of the natural enzyme.

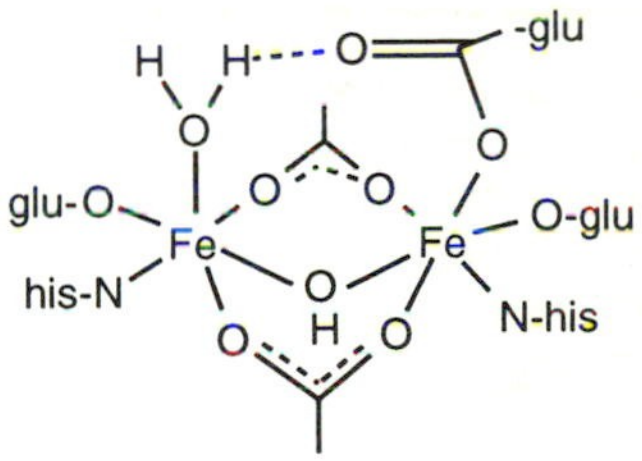

Metal-containing *dioxygenases* usually contain non-haem iron. Among the reactions they catalyse is the oxidative cleavage of catechols (1,2-dihydroxybenzene derivatives) to unsaturated dicarboxylic acids. The structure of one enzyme which has this function has established that the Fe^{III} is distorted trigonal bipyramidal, with axial tyrosine (O) and histidine (N) ligands, and a tyrosine, histidine and water molecule in the equatorial positions. A reactive functional model compound contains tris(2-pyridylmethyl)amine (four N donor atoms) and 3,5-t-butylcatecholate bound to Fe^{III} (Fig. 6.12). The structure is distorted octahedral and the Fe–O bonds are unusually long. The complex causes oxidative cleavage of catechols in the presence of O_2.

Peroxide and superoxide ions, and hydroxyl radicals, even when produced *in vivo*, are highly toxic. Superoxide may be implicated in arthritis,

Fig. 6.11 Core coordination of a methane mono-oxygenase

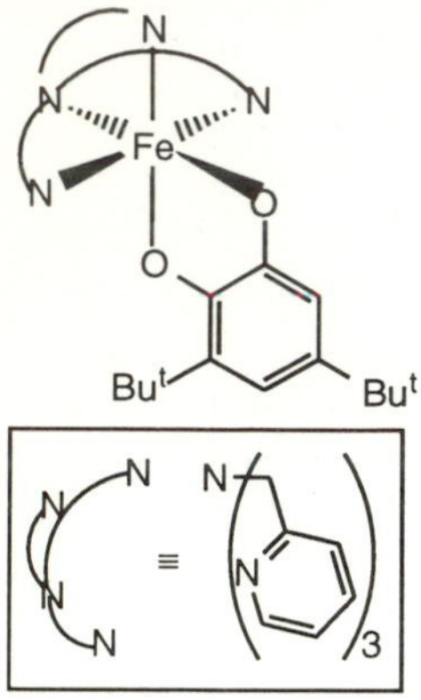

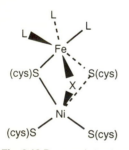

Fig. 6.12 An iron complex which causes oxidative cleavage of catechols

Alzheimer's disease, and in ageing processes. Both peroxide and superoxide may also produce hydroxyl radicals which abstract H atoms from organic molecules and so impair cellular function, modifying DNA bases which can result in mutation. These species must be controlled by appropriate antioxidants, among them the *superoxide dismutases*, *catalases* and *peroxidases*.

Superoxide dismutases can exist either as mononuclear or dinuclear metalloenzymes. The structures of mononuclear Fe and Mn species have been established crystallographically. They have similarities: both have distorted quasi-trigonal bipyramidal geometries with one axial and two equatorial histidine groups and one tyrosine residue. The Mn species has an axial water molecule but the Fe species is four-coordinate with a missing apical ligand, suggesting an entatic state. The mechanism of action of these dismutases is not entirely clear, but the mononuclear iron species may proceed as shown in the following equations:

$$Fe^{III} + O_2^- \rightarrow Fe^{II} + O_2$$

$$Fe^{II} + O_2^- + 2H^+ \rightarrow Fe^{III} + H_2O_2$$

One of the dinuclear superoxide dismutases contains one Cu and one Zn atom which are connected by an imidazolate bridge from a histidinyl residue. The zinc is tetrahedrally coordinated, by one N atom of the bridge, two other histidine groups and an aspartate residue (carboxylate O). The copper(II) centre is distorted square pyramidal with the bridging N imidazole atom in the apical site and the basal sites occupied by three histidinyl residues and a water molecule. The copper ion is the active centre, the water molecule being replaced by superoxide ion.

Peroxidases and *catalases* catalyse the destruction of peroxide. The active centre in *horse radish peroxidase* is an Fe^{III} haem centre (protoporphyrin IX), in which the fifth coordination site is occupied by a histidinyl N atom, and the sixth may either contain H_2O or be vacant. The iron(III) is low spin at high pH and high spin at low pH. Reaction with peroxide generates a ferryl species.

Hydrogenases catalyse the interconversion of dihydrogen and protons. All known hydrogenases contain iron–sulphur clusters, and many also contain nickel, possibly in the form of Ni^{III}. A recent structural determination has shown that one nickel–iron hydrogenase contains two Fe_4S_4 and one Fe_3S_4 cluster, the latter similar to the cubane-like Fe_4S_4 species with one Fe atom missing. These clusters are clearly involved in electron transfer processes, and there is also a hydrogen-binding centre which may involve a heterodinuclear Fe–Ni system. The nickel coordination sphere involves four protein-derived cysteine S atoms, two of which act as bridges to the iron atom which, in turn, appears to carry three or four non-protein ligands L, possibly CO or CN^-. There is an additional bridging group X, possibly O (Fig. 6.13).

Nitrogenases are enzymes which 'fix' atmospheric dinitrogen, *i.e.*, they catalyse the reduction of N_2 to NH_3 at normal temperatures and pressures. This biological process contrasts dramatically with the Haber–Bosch process which achieves the same reduction using solid iron catalysts promoted by

Fig. 6.13 Proposed structure of the active site in a hydrogenase obtained from *Desulfovibrio gigas*

non-transition metal oxides but under relatively high temperatures (*ca.* 450°C) and pressures (*ca.* 270 atm).

The Fe/Mo nitrogenases consist of two proteins. The first, which has a molecular weight of *ca.* 60000, contains one Fe_4S_4 cluster, and the second, with a molecular weight of *ca.* 220000, contains two Mo atoms, 30–32 Fe atoms and approximately the same number of 'inorganic' (non-cysteinyl) S atoms. Nitrogenase obtained from *A.vinelandii* contains two Fe/Mo cofactors based on a Fe_7MoS_8X (X may be O or N) core (Fig. 6.14(a)) and two so-called 'P-clusters', based on linked Fe_4S_4 clusters (Fig. 6.14(b)). The cofactor is bound to the protein superstructur *via* a histidine attached to the Mo (six-coordinate) and a cysteine bound to the Fe (four-coordinate) at the other end of the cluster. What is not known is where N_2 binds in the activation process: it could be at either the Fe or Mo and there are ample models of N_2 complexes of both metals. Other types of nitrogenases are known which contain V in place of Mo, and there is even a system containing only Fe as the transition metal.

NO synthases catalyse the formation of NO from L-arginine in a five-electron oxidation. Nitrogen monoxide is a vital biological messenger, causing muscles to dilate and relax, helping to lower blood pressure, and participating in gut contraction and food movement. It also kills cells and inhibits cell multiplication and growth. All NO synthases use iron protoporphyrin IX (Fig. 6.8(a)) and a number of other coenzymes, and it is well established that NO forms complexes with iron porphyrinato complexes.

Vitamin B_{12} is the human body's defence against pernicious anaemia. It cannot be synthesised in the body and so must be supplied in the diet, and is stored in the liver. The coenzyme form of vitamin B_{12} is based on a cobalt(III) corrin ring complex, containing a Co–C bond. The corrin ring is similar to a porphyrin ring with one –CH= group missing, and provides a near-planar set of four N atoms to support the six-coordinate metal (Fig. 6.15). One axial position is occupied by a 5,6-dimethylbenzimidazole group which is attached, *via* a monosaccharide group linked to a phosphate residue, to the corrin framework. The basic five-coordinate corrin–benzimidazole–cobalt group is often referred to as 'cobalamin', and if the sixth position is occupied by CH_3 the compound is called methylcobalamin (R = Me in Fig. 6.15(a)). The B_{12} coenzyme is 5'-deoxyadenosylcobalamin.

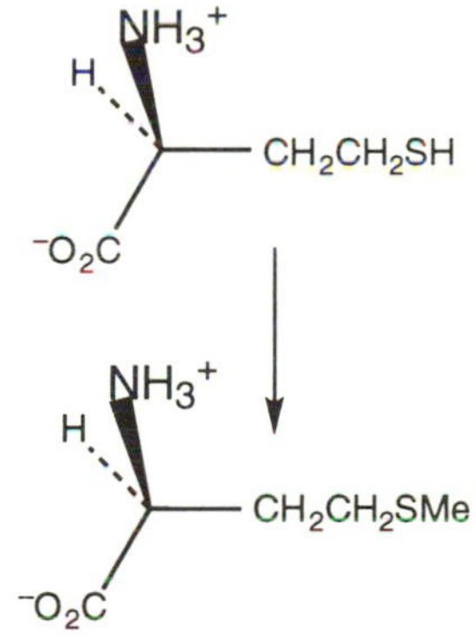

Methylcobalamin is one of three coenzymes involved in the enzyme system *methionine synthase* which catalyses the conversion of homocysteine to methionine (Scheme 6.5). This process appears to involve a reduced cobalamin intermediate, B_{12s}, which abstracts a methyl group from methyltetrahydrofolate (another cofactor), forms a Co–CH_3 bond, and then transfers this methyl group to homocysteine, thereby forming methionine.

A further reaction in which vitamin B_{12} derivatives, particularly methylcobalamin, are implicated is the biomethylation of mercury. Methyl mercury compounds have a catastrophic effect on the brain and central nervous system and are highly toxic as they accumulate faster and remain for longer in the human body than metallic mercury or inorganic mercury salts.

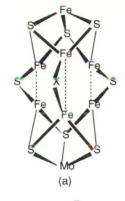

Fig. 6.14 Nitrogenase cofactor: (a) Fe_7MoS_8X and (b) the 'P cluster'

Scheme 6.5 Conversion of homocysteine to methionine

Enzymatic transfer of a methyl group from methylcobalamin to Hg^{2+} salts has been demonstrated.

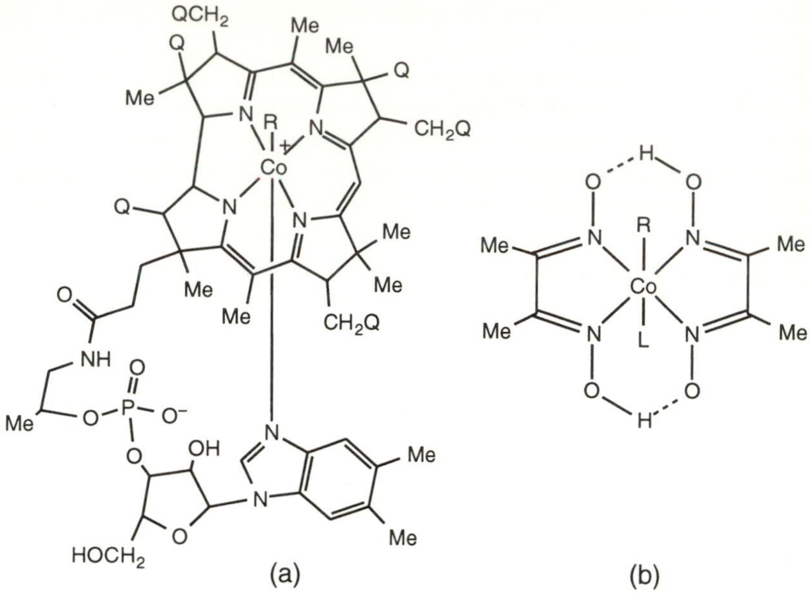

Fig. 6.15 (a) The corrin ring and simplified structure of Vitamin B_{12} (Q = CH_2CONH_2); (b) model cobalt dimethylglyoximato complexes

A considerable number of relatively simple complexes which model the behaviour of vitamin B_{12} have been prepared. The simplest are those based on bis(dimethylglyoximato)cobalt(II) (Fig. 6.14(b)), and cobalt(III) alkyl and other derivatives of this species have been characterised.

6.3 Metals in medicine

Wilson's disease, which leads to dementia and eventual death, is caused by an over-accumulation of copper in the liver and brain. Early chemotherapy exploited the high stability of copper(II) complexes of $EDTA^{4-}$ and related polydentate carboxylic acid salts. Although the copper levels in patients could be lowered relatively efficiently, treatment with this agent also resulted in bone weakening caused by leaching of calcium, a consequence of the high stability of Ca^{2+} complexes of $EDTA^{4-}$. Treatment for this disease is now based on the administration of D-penicillamine (3,3-dimethyl-D-(−)-cysteine, $Me_2C(SH)CH(NH_2)CO_2H$) or its *N*-acetyl derivative (the L-form is toxic). The 'ligand' presents a slightly 'softer' donor atom set (N/O/S) which allows effective discrimination between copper and calcium: copper penicillamine complexes are more thermodynamically stable than those of calcium.

The problem of iron overload in human blood is dealt with by the administration of the siderophore *desferrioxamine B* (Fig. 6.16), a hydroxamato derivative mentioned above on p.81. This species contains three hydroxamato groups which can bind to iron, forming extremely stable complexes which are relatively easily excreted.

The process of administering drugs as 'ligands' to patients is sometimes known as chelation therapy.

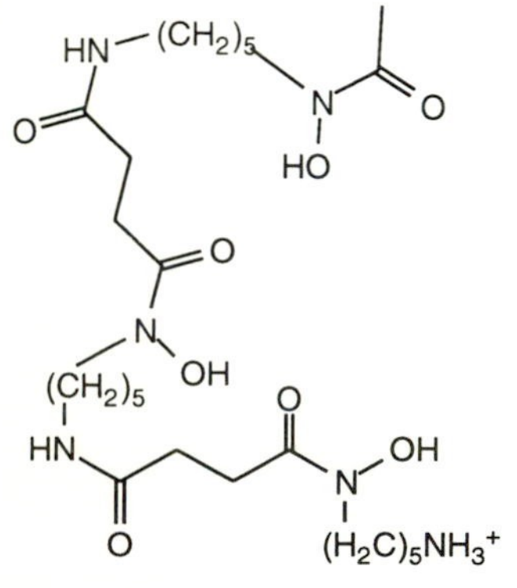

Fig. 6.16 Representation of desferrioxamine B

Further reading

Useful basic Primers:

(a) *d-Block chemistry*, M. J. Winter, Oxford Chemistry Primers 27, OUP, 1994.

(b) *Essentials of inorganic chemistry 1*, D. M. P. Mingos, Oxford Chemistry Primers 28, OUP, 1995.

(c) *The mechanisms of reactions at transition metal sites*, R. A. Henderson, Oxford Chemistry Primers 10, OUP, 1995.

More detailed Primers:

(d) *Biocoordination chemistry*, D. E. Fenton, Oxford Chemistry Primers 25, OUP, 1995.

(e) *Inorganic chemistry in biology*, P. C. Wilkins and R. G. Wilkins, Oxford Chemistry Primers 46, OUP, 1997.

(f) *Organometallics 1*, M. Bochmann, Oxford Chemistry Primers 12, OUP, 1994.

(g) *Organometallics 2*, M. Bochmann, Oxford Chemistry Primers 13, OUP, 1994.

General textbooks:

Advanced inorganic chemistry, F. A. Cotton and G. Wilkinson, 5th Ed., John Wiley & Sons, New York, 1988.

Basic inorganic chemistry, F. A. Cotton, G. Wilkinson, and P. L. Gaus, 2nd Ed., John Wiley & Sons, New York, 1987.

Chemistry of the elements, N. N. Greenwood and A. Earnshaw, Pergamon Press, Oxford, 1984.

Inorganic chemistry, D. F. Shriver, P. W. Atkins, and C. H. Langford, 2nd Ed., Oxford University Press, Oxford, UK, 1994

Inorganic chemistry - principles of structure and reactivity, J. E. Huheey, 3rd Ed., Harper International SI Edition, New York, 1988.

Inorganic structural chemistry, U. Müller, John Wiley & Sons, New York, 1992.

Physical inorganic chemistry, S. F. A. Kettle, Spektrum, Oxford, 1996.

Organometallics, Ch. Elschenbroich and A. Salzer, 2nd Ed., VCH Weinheim, 1992.

Transition metal chemistry - the valence shell in d-block chemistry, M. Gerloch and E. C. Constable, VCH Weinheim, 1994.

Index